C.E.T.A

Close Engagement Target Acquisition

Dr. Leonard C. Holifield, DMS

Copyright Notice

Published by Kinetic Digital Publishers

www.kineticdigitalpublishers.com

For permissions, inquiries, or other correspondence, please visit our website.

DEDICATION

This book is dedicated to the men and women of the armed forces, law enforcement, executive protection agents, security officers, special operation units, martial artist and those civilians seeking real-world knowledge in defending against a life-threatening attack. This book is also dedicated to my parents Cleve and Ruth Holifield to whom this work would not have been possible without. To CSM Lesley Freeman Jr, whom I served under while stationed at Camp Casey, Korea, 2nd Infantry Division and who believed in me and my mission while training troops in hand-to-hand combat. To Col. Dennis R. Lewis, whom I served under at Ft. Bragg, NC. and who also believed in my mission to enhance close quarter combat training for the battalion. To SMA Richard A. Kidd, whom I met with at Ft. Sill, OK, FATC - Field Artillery Training Center to discuss the future of hand-to-hand combat in today's Army. To my lead assistant instructor James E. Turner who later retired SGM, Special Forces, and my CQC demo team in Korea for your support and dedication to me and our mission to revitalize hand-to-hand combat. There are literally too many names to mention here who supported me during my military career in advancing hand-to-hand combat for our men and women in uniform; I thank you for your support. Last but not least, all thanks to God, for without the most-high, I would not exist to bring this work to you.

DISCLAIMER

This textbook offers an in-depth, scientific examination of the human body's vulnerabilities, specifically focusing on methods to disable or neutralize an individual by targeting key life support systems. Developed from the principles of C.E.T.A. (Close-Engagement-Target-Acquisition), this work provides a detailed analysis of the Central Nervous System, Circulatory System, Respiratory System, Brain, Skeletal System, and Internal Organs. Each section includes medical insights into the damage inflicted and its physiological consequences, presented for academic study.

Tailored for military personnel, special operations units, law enforcement, martial artists, bodyguards, and executive protection agents, this text is intended for professionals seeking a comprehensive understanding of human biomechanics in high-threat situations. It delves into the science behind incapacitation, offering a tactical perspective on physical confrontation. The material covered in this book is strictly for academic study and educational purposes only. The reader is solely responsible for the use or misuse of the information provided in this book and the author is not responsible for any damage done as a result of the use of this information.

UNITED STATES ARMY
THE SERGEANT MAJOR

July 11, 2023

Dr. Leonard Holifield

Dear Dr. Holifield:

I want to thank you for the outstanding and unyielding support you have provided to the United States Army and our Soldiers with your contributions in the field of Army Combatives and as Chief Combatives Instructor. Your professionalism, selfless service, and dedication marks you as a true leader and reflects the ultimate military professionalism. Thank you for your exceptional service for so many years and best of luck in your future.

Once again, thank you for all that you have done for our Army and our great Nation. Be All You Can Be!

Sincerely,

Michael A. Grinston
Sergeant Major of the Army

TABLE OF CONTENTS

CLOSE QUARTERS COMBAT
CHIEF
INSTRUCTOR
U.S.ARMY
C.E.T.A.

CHAPTER 1

INTRODUCTION TO C.E.T.A.

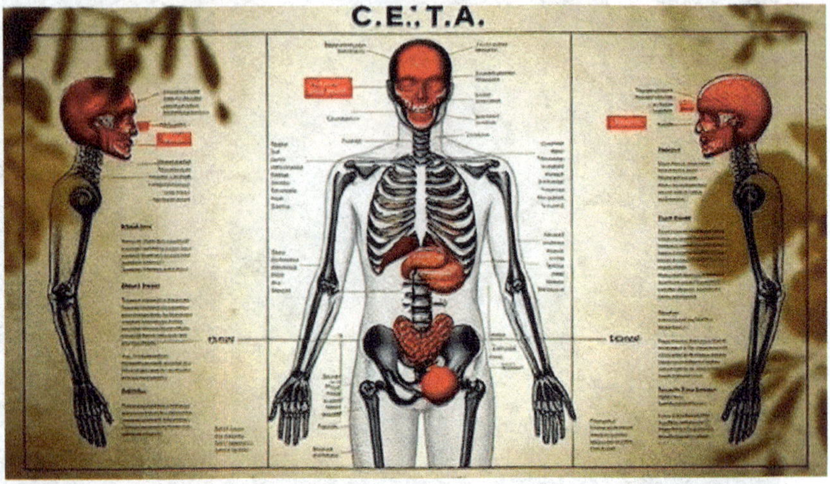

Defining Close Engagement Target Acquisition (C.E.T.A.)

Close-Engagement-Target-Acquisition, or C.E.T.A., represents a specialized system engineered for the rapid and decisive neutralization of immediate threats. Developed from extensive experience in close-quarters combat and tactical training, C.E.T.A. moves beyond conventional self-defense by focusing on the physiological vulnerabilities of the human body. Its core principle centers on ending physical confrontations with maximum efficiency and minimal wasted effort. This approach is rooted in a deep understanding of human anatomy and biomechanics, providing a systematic framework for critical engagement.

Unlike general combative, C.E.T.A. systematically identifies and exploits the body's essential life support systems as primary targets. This includes the Central Nervous System, the intricate

Circulatory and Respiratory systems, the brain, and crucial internal organs. The methodology involves a calculated analysis of these areas, aiming to disable or incapacitate an aggressor through precise, targeted strikes. Understanding the medical consequences of such damage forms a critical component of this tactical science.

The objective of C.E.T.A. is not merely to defend, but to decisively conclude a high-threat encounter by rendering an opponent nonfunctional. This requires a shift from reactive techniques to a proactive, analytical mindset, viewing the human body as a collection of systems with exploitable weaknesses. For professionals operating in high-stakes environments, C.E.T.A. offers a rigorous, scientifically-backed approach to threat neutralization, prioritizing tactical effectiveness above all else. Its application demands a comprehensive understanding of both physiological impact and professional responsibility.

The Scientific Basis of Tactical Neutralization

C.E.T.A. is fundamentally built upon a deep, scientific understanding of human anatomy and physiology. It moves beyond the simplistic application of force, instead advocating for precise targeting based on known physiological vulnerabilities. This approach ensures that actions taken are not random strikes, but calculated applications designed to disrupt critical bodily functions. Such precision is paramount for achieving rapid and decisive threat neutralization.

To truly incapacitate an individual, one must comprehend the intricate network of the Central Nervous System, the vital flow of the Circulatory System, and the essential mechanics of the Respiratory System. Each of these systems represents a potential point of failure, a vulnerability that, when exploited with accurate knowledge, can lead to immediate and predictable outcomes. This detailed physiological insight transforms a chaotic confrontation into a controlled, tactical engagement, minimizing collateral and maximizing efficiency. It is the application of medical understanding to the grim realities of close-quarters combat.

Scope and Applications of C.E.T.A.

The principles of C.E.T.A. are engineered for direct application across a spectrum of high-stakes professions where rapid, decisive action is paramount. Military personnel, special operations units, and law enforcement officers gain a critical framework for effective threat neutralization in dynamic and unpredictable environments. This system provides a scientifically grounded approach to incapacitation, moving beyond traditional brute force to a more calculated, anatomically precise methodology. Its utility extends to situations demanding swift resolution, minimizing collateral damage while maximizing operational effectiveness.

For martial artists, bodyguards, and executive protection agents, C.E.T.A. offers an advanced understanding of physiological vulnerabilities, enhancing their defensive and offensive capabilities. This knowledge allows for the precise targeting of critical life support systems, ensuring that engagements are concluded with optimal efficiency and control. The focus shifts from merely engaging an adversary to strategically dismantling their capacity for further aggression. Ultimately, C.E.T.A. equips these professionals with the most direct and effective means to manage and terminate immediate physical threats, safeguarding lives and maintaining security.

Ethical Considerations and Professional Responsibility

The profound insights offered within C.E.T.A. concerning the human body's vulnerabilities carry an immense professional and ethical weight. This text, designed for rigorous academic study and specialized tactical application, demands a deep understanding of its inherent responsibilities. Readers are expected to approach this material with the utmost seriousness, recognizing the gravity of the knowledge being imparted.

Applying the principles outlined herein requires unwavering adherence to established legal frameworks, rules of engagement, and moral codes pertinent to one's operational domain. The information is strictly intended for scenarios demanding the rapid and effective neutralization of a threat, typically within the confines of self-defense or authorized professional duty. Misuse or casual application of these techniques carries severe, irreversible consequences, both legally and ethically.

Therefore, the onus of responsible implementation falls squarely upon the individual practitioner. Dr. Holifield's work serves as a comprehensive guide for those operating in high-stakes environments, not a manual for wanton aggression. Mastery of C.E.T.A. is synonymous with a profound commitment to disciplined, ethical conduct in the face of extreme threats.

Combat Psychology

THE AGGRESSIVE MINDSET

An aggressive mindset puts you on the offensive and your opponent on the defensive. The important thing here is that you continue to press them. Attack, making it difficult for your opponent to counterattack. You must be aggressive and unyielding in your attack towards your opponent. An aggressive mindset can be intimidating to your opponent, thus causing him to lose confidence in his own ability to defeat you.

MENTAL CONDITIONING

The outcome of any physical confrontation will depend solely on your mental state of mind going into the fight. You must develop an attitude and state of mind that will allow you to engage with your opponent without hesitation. For fear of personal injury. Period. To accomplish this, you must first separate yourself from the fact that your opponent is Human. And instead. Look at him. As a target, not as a human being, but a target that must be destroyed! You must also have the utmost confidence in your techniques and your ability to use them effectively. Most traditional martial arts systems teach literally hundreds. Of self-defense techniques, which can and have caused information overload. During the initial stages of confrontation. The brain starts to try to sort through tons of information or techniques. That it has stored through the years for specific situations, By the time the brain finds the technique best suited for the situation, the fight is over, and you are just starting to gain consciousness from being knocked out several minutes ago. The brain simply cannot compute. Or nor process that much information and time to react with the proper technique at the precise time of the attack or strike. Learned techniques require thought. The time elapsed between thought and reaction can literally mean the difference between life and death. Fights to the death during combat are almost over as fast as they start. The action is fast and furious, and there is little or no time to think about what you're going to do and how? They're going to do it! Therefore, it is best to learn only a small number of techniques, two or three for a specific attack. The fewer techniques, the less time the brain has to find the technique, and the faster you can react to the attack. This is a mental condition called unconsciously competent. Where the technique becomes instinctive, second nature, and instantaneous counterattack. This will be covered in more detail later in this book. A thorough knowledge of human anatomy and vital target points of the body is essential in building your confidence level, which again will allow you to engage the enemy without fear or hesitation. Remember, in situations where time is of the essence, less is best.

CONTROLLING FEAR AND ANGER

Fear and anger can both affect you in different ways, therefore, it is important to understand them both. Fear is a healthy defense mechanism built into the human psyche, which can protect you from injury, especially when confronted with situations beyond your control. This fear can have both a negative or positive effect on you, depending on how you deal with it. Fear triggers certain chemical reactions to take place in the body. One of these reactions is an increase in adrenaline, which. Often causes an increase in physical strength, speed, and power. On the other hand, the same adrenal rush can also cause you to hesitate and freeze altogether. Which gives a great deal of truth to the old saying, fight, flight, or freeze. The bottom line here is that fear can be controlled and used to your advantage. This goes back to mental conditioning. The more confident you are in yourself, the less likely you are to be afraid to engage. Fear usually comes from not knowing. What is on the other side, or what's about to happen? The more you train for these types of scenarios, the better prepared you will be to meet them head on. Anger is an emotion that can easily be controlled. Again, as I mentioned earlier, the more prepared you are, the less likely you are to get angry and let your emotions take you over. Anger can be your worst enemy and your downfall as well if you don't learn to control your temper. You never want your emotions to act without the benefit of intellect and or logic. I have been involved in the martial arts for over 50 years. And during that time, I've performed many breaking techniques. To Tameshiwari, the art of breaking, for audiences around the world. Many were successful, and some unsuccessful. I use what I call the baseball theory: Three strikes and you're out. If my first attempt was unsuccessful, I would try a second time. If unsuccessful, on the 2nd, I would try 1/3 time, If unsuccessful on the third try, I would graciously bow out. You see, before your first attempt. You are mentally alert, competent, and ready. After the first failure, you lose a bit of confidence, focus, relaxation, and tend to be a little more tense, after the 2nd. Failure, Frustration starts to set in, and your pride is hit hard, after the third failure, anger sets in, and the focus and concentration is greatly reduced in any attempt to break at this time. May and has resulted in broken bones, humiliation, embarrassment, and sometimes death, depending on the type of break, IE, head breaks, for example. Therefore, it is imperative that you stay relaxed, focused, and in control. In your situation, no matter what it may be, you will find that anger can be controlled and used to your advantage.

THE FOUR LEVELS OF CONSCIOUSNESS

There are 4 levels of consciousness that play a vital role in your ability to understand your own personal level of competency and that of your opponent.

1. UNCONSCIOUSLY INCOMPETENT

At this beginning stage, you are literally at Ground Zero. A state of mind. Where you have no idea what is going on. Ignorant of the subject matter. A beginner, novice, or rookie.

2. CONSCIOUSLY INCOMPETENT

Having no realistic competency of skills taught. A state of mind where you know that you are lacking skill.

3. CONSCIOUSLY COMPETENT

At this stage, you know the skill or technique and can execute the same, however. You need time to think before reacting. Most martial artists fall into this category and stay there. It should be your objective to move forward to the point where you no longer have to think. About how you will react, but act!

4. UNCONSCIOUSLY COMPETENT

The final level. The level of second nature. Your reaction time and senses have sharpened and increased to the point where your reactions have become automatic. Or second nature. This is the ultimate level that you should strive for.

THE TEN BASIC CETA PRINCIPLES

BALANCE

Proper weight distribution, foot placement, and stability.

CONTROL

Controlling your fears, anger, and emotions.

POSITIONING

Positioning your body in relation to your attackers using distance and angles.

TIMING

Using sound judgment, not act too soon. Or too late, but exactly at the right time.

DISTANCE

The space created between you and your opponent.

MOMENTUM

Using your opponent's force or momentum against him, to your advantage.

LEVERAGE

Using the elbow, wrist, or knee as a fulcrum or pivot point. To control your opponent.

SPEED

Speed is essential in generating power as well as getting in and getting out fast.

COMMITMENT

To commit is to give yourself to the enemy without fear of injury or death. When you do not fear death, you have nothing else to lose, and as a result, you become fierce, unyielding, and unstoppable in your attack and execution.

EXECUTION

You must pay close attention to detail and the mechanics of each technique to ensure proper execution of the technique.

THE FOUR MAJOR CETA PRINCIPLES

CLEAR

To move your body out of the line of attack. When defending against an attack, you must move your body off the line of attack to avoid the attack. This can be accomplished by stepping off of the line of attack at various angles or pairing. To the left or to the right.

STABILIZE

To control the weapon, hand, arm, or leg. This can be done by the use of a joint. Manipulation techniques to stabilize the weapon hand, arm, or leg.

DISARM

To relieve the attacker of his weapon. A weapon, IE. , Rifle, Knife, Club, Gun, etc, can make anyone a formidable opponent. You must disarm the attacker to lessen the threat level. This can be done in a variety of ways, However, you must keep in mind that if the weapon is active in one hand, this does not mean that the other hand is on vacation. Period. You must be aware of your entire situation and surroundings. And that your attacker may have a secondary weapon as well. Therefore, you must be able to act quickly and decisively to disarm, stabilize, and neutralize the attacker. Another way to lessen the threat by disarming the attacker of a firearm is to direct the muzzle of the gun to the ground or into the sky. Caused the attacker to fire his shots away from you. Your objective here is to empty the gun. Remember, you are less likely to be killed by an empty gun than a loaded one.

NEUTRALIZE

To incapacitate, immobilize, or kill your attacker. Virtualization can be accomplished by 1 restraining, 2 immobilizations of the skeletal support system by the breaking of the bones, 3 rendering attacker unconscious, or 4 killing the attacker.

THE FOUR RANGES OF ENGAGEMENT

LONG RANGE

Long range engagement involves the use of the legs to employ various kicks and sweeps, due to the reach of these weapons and the distance it places between you and your attacker.

MEDIUM RANGE

Medium range engagement involves the use of the hands and the arms in punching and striking.

CLOSE RANGE

Close range engagement involves the use of close quarter trapping techniques. These techniques employ the use of blocks and simultaneous strikes.

WRESTLING RANGE

The fourth and final stage of engagement. Involve the use of wrestling or grappling techniques. These techniques are usually employed after the fight has gone from a standing position. To the ground.

C.E.T.A CONCEPTS

FORCE REDIRECTION

Force redirection. Focuses on the redirection of your attacker's movement and force towards you, in another direction, as opposed to using force against force in meeting the attack. As the attacker attacks, you move out of the line of attack. By pairing to the left or to the right or at a 45° angle to the attack, allowing the attackers strike. And forced to go past you.

CLEARING THE LINE Of ATTACK

This concept goes hand in hand with force redirection, by moving your body or target area, arm, leg, head, etcetera, out of the way of attack, thereby preventing you from being hit.

CONTINUOUS ATTACK

This concept involves. Pushing the attack on your opponent, by overpowering him with multiple offensive strikes, kicks, etcetera. By continually attacking your opponent. This places him on the defensive, where he must use blocks to avoid. Being hit, which places you at a great advantage to land several disabling blows, while being on the offensive.

ECONOMY Of MOTION

This is the process of using only what is necessary in regards to body mechanics to subdue the attacker as possible.

RELAXATION

The more relaxed you are, the less stressed you will be, which will allow you to think clearly and execute your techniques.

SPEED / POWER

Speed equals' power, and power equals mass times velocity (MxV=P). To have maximum speed, you must first be relaxed. Remember that the more relaxed you are mentally and physically, the more speed you will generate, and the more speed you generate, the more power you will have going into the target period.

ENGAGEMENT STRATEGY

TELEGRAPHING

Telegraphing is the process of showing your intent. To execute a particular move or technique, whereby giving your opponent a preview, if you will, of what you are about to do. This will most certainly set you up for failure, as your opponent will now have a counter move waiting for you. Never Telegraph. You are about to do to your opponent, unless you are using the Telegraph as a fake or distraction to set your opponent up for your primary attack.

ANTICIPATING

When you anticipate something about to happen, you tend to focus your thoughts on a specific action that you believe is about to take place. In unarmed self-defense situations, this can almost certainly place you at a great disadvantage at the hands of your attacker. For example, if you anticipate your attacker throwing A right-handed hook punch towards your face, you have subconsciously prepared your mind and body to react to this attack by blocking the incoming punch. Though logical, the problem here is that, instead of your opponent throwing that right hook, it kicks you in the groin instead, catching you completely off guard! Therefore, it is best not to anticipate or second guess your opponent. Combat is unpredictable. And though you may be able to guess what your opponent's move may be, it would be unwise to make this way of thinking a habit.

FLOWING

To help eliminate the possibility of anticipating your opponent's next move, you must be able to allow yourself to flow or blend with your opponent. Right following with your opponent, you make yourself one with him. Focusing not on one specific part. Or the movement of his body, but his entire being. In other words, look at your opponent's eyes, but see everything using your entire field of vision to your advantage. By shadowing or flowing with your opponent, you are in a much better position to meet, defend, and counter whatever attack he launches at you.

DECEPTION

Deception is one of the oldest tricks in the book. To deceive your opponent is to make your opponent believe something that is not true, causing him to change his course of action, to your benefit or advantage. An excellent example of this strategy was used by the legendary boxing great Muhammad Ali. During his fight with George Foreman, Ali, knowing his opponent was much larger and stronger than he was, leaned against the ropes and let form and work on his body, blow after blow, he battered Allie's midsection and ribs. Period. To the thousands of spectators that night, it looked like Ali was in real trouble, however. It was all part of Aldi's master plan of deception. His plan was to wear foreman down to the point of total exhaustion, as Ali knew that he himself was in top physical condition. His plan worked flawlessly as foreman. Lost his strength, power, and stamina. And Ali came back stronger than ever to defeat George

Foreman in one of the greatest upsets in boxing history. You may have heard of this technique. The Rope-a-dope.

PREMEDITATING

Never pre-plan a technique. This borderlines on the same principle as anticipating, however, with a different twist period. To premeditate or preplan A technique, is to tell yourself. Parenthesis. When this guy punches, I'm going to block his punch using an outside block and follow up with a reverse punch to his solar plexus, then a hook punch to his jaw, and a follow up with a one arm shoulder throw. Surprise!!! He doesn't punch at all. Instead, he kicks you in the groin. Get the point?! This goes back to flowing with your opponent.

FORCE REDIRECTION

In most cases, it is much easier to redirect your opponent's force than to meet it head on. This is especially true in hard martial art styles such as Shotokan karate, where hard, powerful linear blocks, kicks, and strikes are used. The traditional martial art of Aikido is based on the concept of force redirection and has been used quite effectively by those practitioners who have mastered its principles and techniques. Forced redirection is simply the process of redirecting your opponent's force, ie, punch, or kick, in a different direction, which is advantageous to you, but places your opponent at a great disadvantage. This can be accomplished by moving your body out of the line of attack, where allowing your attacker's momentum to continue, and redirecting his attack in a position that sets him up for a follow up technique.

CHAPTER 2

THE CENTRAL NERVOUS SYSTEM

COMMAND AND CONTROL

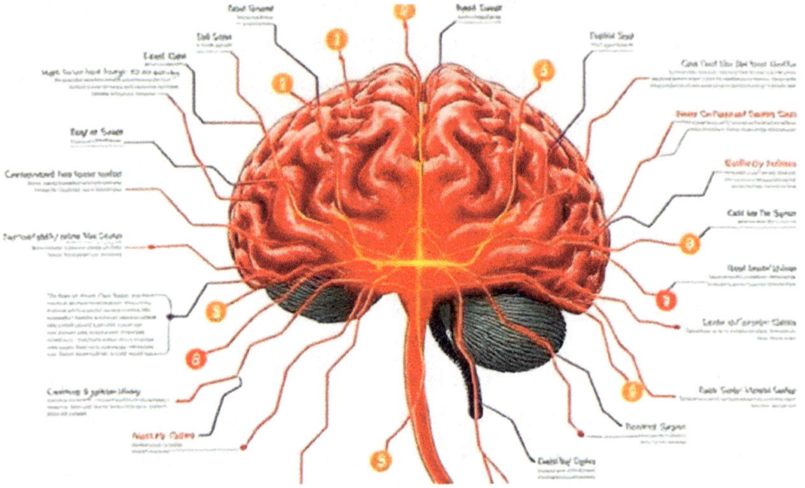

Anatomy and Physiology of the Central Nervous System

The Central Nervous System, or CNS, stands as the unequivocal command and control center for the entire human organism. Comprising the brain and spinal cord, this intricate network orchestrates every conscious thought, involuntary reflex, and vital physiological process. Its integrity is paramount for survival, acting as the master regulator of all bodily functions. Any disruption to its operations directly compromises an individual's capacity for action or even life itself.

The brain, nestled within the cranium, serves as the primary processing unit, interpreting sensory input and issuing motor commands. Descending from it, the spinal cord acts as the critical conduit, transmitting these impulses to and from the peripheral nervous system. This continuous, rapid communication loop is what enables movement, perception, and the maintenance of homeostasis. Understanding this fundamental architecture is the first step in comprehending how specific interventions can effectively dismantle an opponent's operational capabilities.

Vulnerabilities and Critical Nodes

While the central nervous system enjoys robust protection from the skull and vertebral column, it possesses distinct anatomical points where its vulnerability to external kinetic force becomes acutely pronounced. These specific locations, termed "critical nodes," are areas where a precisely applied impact can rapidly compromise the CNS's structural integrity and functional capacity. Identifying and understanding these inherent weaknesses is foundational for any professional seeking to effectively and immediately neutralize a threat.

Such critical nodes are not random; they correspond to anatomical regions where the protective bone or muscular layers are thinnest, or where vital nerve structures are superficial. Examples include the mastoid process behind the ear, which can transmit impact directly to the brainstem, or specific points along the cervical spine where the vertebral column's stability is compromised. These areas offer direct access to disrupt the intricate pathways of neurological control.

Exploiting these critical nodes is less about overwhelming force and more about surgical precision. A well-placed strike to one of these targets can induce immediate neurological dysfunction, ranging from temporary stunning to profound unconsciousness, by disrupting vital signals. This targeted approach aligns perfectly with the C.E.T.A. principle of achieving rapid incapacitation through the efficient compromise of the body's core life support systems.

Mechanisms of CNS Incapacitation

Incapacitation of the Central Nervous System, the body's primary control network, is achieved through various mechanisms, each designed to disrupt neural integrity. Direct blunt force trauma to critical areas, such as the cranium or cervical spine, can induce immediate neurological dysfunction. This impact generates a concussive wave, causing axonal shearing and widespread neuronal depolarization. Such a violent disruption compromises the brain's ability to send and receive vital signals, leading to rapid loss of consciousness or motor control.

Beyond blunt force, sharp force trauma, or extreme compression targeting the spinal cord can sever or crush neural pathways. A complete transection of the spinal cord, for instance, results in immediate and irreversible paralysis below the point of injury, effectively isolating the brain from the rest of the body. Furthermore, mechanisms that induce severe oxygen deprivation to the brain, such as carotid artery compression, rapidly deplete neuronal energy reserves. Without adequate oxygen, neural cells cease to function, leading to swift incapacitation and potentially irreversible damage.

Even without direct structural damage, extreme physiological shock can overwhelm the CNS, leading to a temporary, but profound systemic shutdown. Intense pain stimuli directed at highly innervated areas, or sudden, overwhelming sensory input, can trigger a vasovagal response. This reflex can cause a precipitous drop in heart rate and blood pressure, resulting in cerebral hypoperfusion and subsequent syncope. Such non-destructive mechanisms still achieve the objective of immediate, albeit transient, incapacitation.

Physiological Consequences of CNS Trauma

When the central nervous system suffers trauma, particularly to the brainstem or upper spinal cord, the effects are immediate and catastrophic. Such an impact instantly compromises vital autonomic functions like respiration and cardiac rhythm, leading to rapid systemic collapse. This often results in immediate unconsciousness or a complete loss of motor control, rendering an individual instantly non-functional. The sudden cessation of command signals from the brain effectively shuts down the body's operational capacity, eliminating any sustained threat.

Depending on the precise area impacted, CNS trauma can also manifest in a broader spectrum of neurological deficits. Injury to the cerebrum, for instance, might induce profound cognitive impairment, disorientation, or even a persistent vegetative state, even if basic vital signs temporarily endure. Conversely, damage to the spinal cord immediately severs the critical communication pathways between the brain and the body, resulting in flaccid paralysis below the point of impact. This comprehensive disruption prevents any coordinated movement or response, thereby neutralizing the subject's ability to pose any further aggressive action.

CNS trauma triggers a sequence of progressive and destructive physiological responses that extend beyond the initial mechanical injury. This process, known as secondary injury, can last for hours, days, or even months after the initial event, leading to widespread damage, cell death, and long-term functional deficits.

Primary injury

The immediate, primary injury is the direct mechanical damage to brain or spinal cord tissue at the moment of impact. It is irreversible and includes:

- **Contusions:** Bruises and bleeding on the surface of the brain.

- **Lacerations:** Tearing of brain tissue or blood vessels.
- **Hemorrhage:** Bleeding within the brain tissue (intraparenchymal) or in the surrounding spaces (e.g., subdural, epidural, subarachnoid).
- **Diffuse axonal injury (DAI):** Widespread shearing and stretching of nerve fibers, disrupting communication between neurons.

Secondary injury

Secondary injury is a cascade of cellular and molecular events triggered by the initial trauma. The goal of emergency treatment is to mitigate these progressive and damaging effects.

Cerebral and spinal cord swelling (edema)

- **Vasogenic edema:** The blood-brain barrier (BBB) is disrupted, allowing fluid and proteins to leak from blood vessels into the brain's extracellular space. This swelling, combined with any hemorrhage, increases intracranial pressure (ICP).
- **Cytotoxic edema:** Injury to brain cells impairs their ion pumps. As a result, water rushes into the cells, causing them to swell and potentially rupture.

Increased intracranial pressure (ICP)

The skull is a rigid compartment, so swelling or bleeding increases pressure inside the skull. This elevated ICP reduces cerebral perfusion pressure (CPP), which is the pressure of blood flowing to the brain.

- Reduced blood flow deprives brain tissue of oxygen and nutrients, leading to ischemia and further cell death.
- **Herniation:** A severe consequence of increased ICP,

where the brain tissue is forced out of its normal position, can compress the brainstem and cause death.

- **Cushing's triad:** A clinical sign of dangerously high ICP, characterized by increased blood pressure, a slow heart rate (bradycardia), and irregular breathing.

Excitotoxicity and ion imbalance

- CNS trauma causes neurons to release excessive amounts of the excitatory neurotransmitter glutamate.

This overstimulates glutamate receptors, causing a massive influx of calcium ions (

- Excess intracellular calcium triggers destructive enzymes and impairs mitochondria, a process that can ultimately lead to cell death.

Neuroinflammation

- Glial cells (astrocytes and microglia) and infiltrating immune cells from the bloodstream release pro-inflammatory cytokines and chemokines.
- While initially helpful for cleaning debris, prolonged inflammation can lead to further damage of healthy tissue. Chronic inflammation can persist for months or years after the initial injury.

Oxidative stress

- Trauma and the resulting cascades generate an excessive amount of unstable molecules called free radicals.
- These free radicals overwhelm the body's natural antioxidant defenses and cause irreversible damage to lipids, proteins, and DNA within brain cells.

Mitochondrial dysfunction

- Brain injury disrupts the function of mitochondria, the cells' powerhouses.
- Impaired mitochondrial function exacerbates energy deficits, disrupts calcium regulation, and promotes the release of proteins that trigger cell death.

Vascular and autonomic system dysfunction

- **Cerebrovascular autoregulation:** The brain's ability to maintain a consistent blood flow despite changes in blood pressure can be impaired, making it vulnerable to drops in blood pressure.
- **Spinal cord hemorrhage and ischemia:** In spinal cord injury, hemorrhage and a drastic reduction in blood flow can cause cell death and worsen neurological deficits.
- **Autonomic nervous system (ANS) instability:** Especially after traumatic brain injury (TBI) or high-level spinal cord injury, patients may experience paroxysmal sympathetic hyperactivity, characterized by exaggerated vital signs such as high blood pressure, rapid heart rate, and hyperthermia. This is often associated with dysregulation of the hypothalamus.
- **Cerebrovascular autoregulation:** The brain's ability to maintain a consistent blood flow despite changes in blood pressure can be impaired, making it vulnerable to drops in blood pressure.
- **Spinal cord hemorrhage and ischemia:** In spinal cord injury, hemorrhage and a drastic reduction in blood flow can cause cell death and worsen neurological deficits.
- **Autonomic nervous system (ANS) instability:** Especially after traumatic brain injury (TBI) or high-level

spinal cord injury, patients may experience paroxysmal sympathetic hyperactivity, characterized by exaggerated vital signs such as high blood pressure, rapid heart rate, and hyperthermia. This is often associated with dysregulation of the hypothalamus.

CHAPTER 3

THE CIRCULATORY SYSTEM

THE LIFEBLOOD

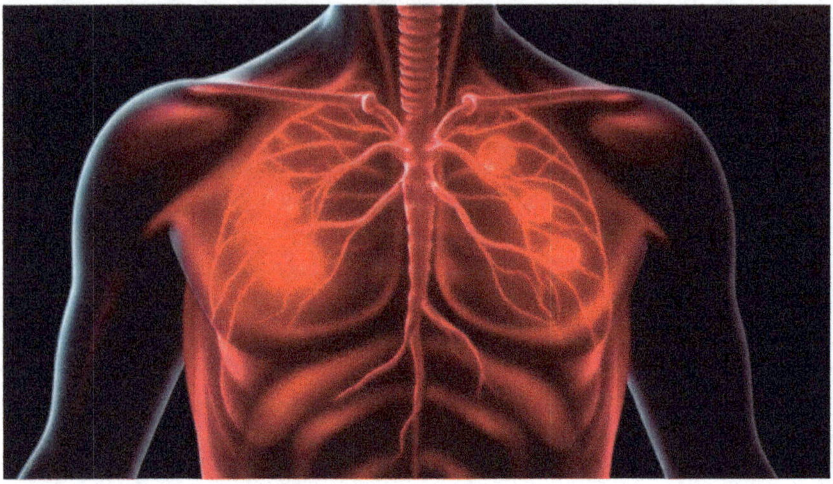

Anatomy and Physiology of the Circulatory System

The circulatory system stands as the body's indispensable transport network, a complex arrangement of the heart, blood vessels, and blood itself. Its primary function is to deliver oxygen, nutrients, hormones, and other vital substances to every cell while simultaneously removing metabolic waste products. This intricate vascular tree, comprising arteries, veins, and capillaries, ensures constant cellular respiration and overall physiological equilibrium. Without this continuous, unimpeded flow, the body's delicate internal environment rapidly destabilizes.

Functioning as a closed hydraulic circuit, the circulatory system maintains a precise balance of pressure and volume essential for life. The heart acts as the central pump, propelling blood through miles of vessels, each designed for specific roles in distribution and collection. Any compromise to this system, whether through direct trauma, internal hemorrhage, or systemic shock, critically impairs the body's ability to sustain vital functions. Understanding these foundational mechanics is paramount for comprehending how targeted disruption can lead to rapid incapacitation.

Key Vulnerabilities and Hemorrhage Points

Understanding the intricate architecture of the circulatory system is only the foundational step; true tactical proficiency lies in identifying its critical failure points. These specific "hemorrhage points" represent anatomical locations where major blood vessels are either superficially accessible or vulnerable due to their proximity to bone and nerve clusters. Exploiting these precise sites allows for rapid, decisive incapacitation through acute blood loss or systemic circulatory collapse. Mastering this knowledge is therefore paramount for efficient threat neutralization, minimizing engagement time, and maximizing effectiveness.

Several key arterial pathways stand out as primary targets due to their significant blood flow and relative exposure. The carotid arteries, located bilaterally in the neck, are particularly vulnerable as they are large, superficial, and directly supply the brain, making them prime targets for both compression and laceration. Further, the femoral artery, situated in the groin, presents another significant vulnerability due to its substantial blood volume and relatively exposed position, especially during dynamic engagements. Similarly, the brachial artery in the inner arm and the subclavian artery beneath the clavicle offer critical points for disruption.

Compromising these major arterial and venous pathways initiates an immediate and cascading physiological crisis within the human body. Rapid external or internal hemorrhage quickly deprives the brain and vital organs of essential oxygen and nutrients, leading to a swift onset of hypovolemic shock, loss of consciousness, and ultimately, systemic failure. Even partial compromise can severely impair an adversary's motor control and cognitive function, rendering them unable to continue an assault. The objective is not merely to inflict damage, but to induce an irreversible physiological cascade that neutralizes the threat decisively.

Disrupting Circulation: Tactical Implications

Disrupting the circulatory system is not merely about causing injury; it is a calculated act aimed at achieving immediate and decisive incapacitation. By compromising the body's primary transport mechanism for oxygen and nutrients, an assailant can induce rapid physiological collapse. This tactical objective aligns directly with C.E.T.A. principles, focusing on swift neutralization by targeting a critical life support system. The resulting loss of consciousness and motor control offers a definitive end to a high-threat encounter.

Effective circulatory disruption necessitates precise targeting of major arterial or venous structures, particularly those identified as key hemorrhage points. Direct trauma, compression, or severing actions applied to these vulnerable areas can precipitate a rapid and catastrophic loss of blood pressure and cerebral perfusion. Such targeted interventions are designed to overwhelm the body's compensatory mechanisms, leading to a swift and irreversible decline in combat effectiveness. The objective is not merely to wound, but to dismantle the opponent's capacity for continued aggression.

Physiological Effects of Circulatory Compromise

Compromising the circulatory system initiates a rapid cascade of severe physiological events. The immediate consequence of significant vascular damage is rapid blood loss, leading to hypovolemic shock. As blood volume diminishes, the heart struggles to maintain adequate perfusion, resulting in a critical reduction of oxygen and nutrient delivery to vital tissues. This deprivation quickly impairs cellular function, particularly in the brain and major organs, signaling the onset of systemic failure.

With continued circulatory compromise, the body's compensatory mechanisms rapidly become overwhelmed. Oxygen starvation, or hypoxia, leads to metabolic acidosis as cells switch to anaerobic respiration, further poisoning the system. Brain function deteriorates rapidly, manifesting as disorientation, loss of consciousness, and eventual cessation of neural activity. Simultaneously, critical organs like the kidneys and liver begin to fail, unable to process waste or maintain homeostasis without adequate blood flow.

Ultimately, the profound physiological effects of circulatory compromise culminate in systemic collapse and death. The body's intricate network of life support systems, deprived of their essential fuel and waste removal, simply cannot sustain function. Understanding this progression is paramount for tactical professionals, as it provides the scientific basis for incapacitation strategies targeting the circulatory system. This knowledge underscores the critical importance of rapid and decisive action in high-threat scenarios.

CHAPTER 4

THE RESPIRATORY SYSTEM

THE BREATH OF LIFE

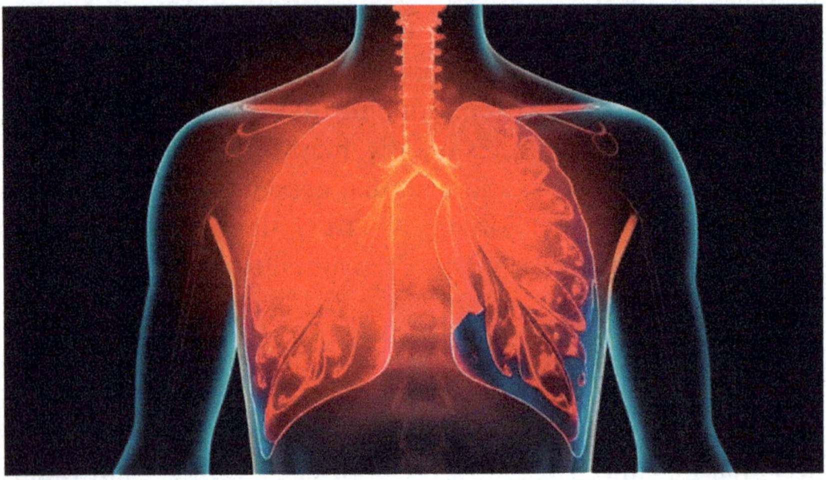

Anatomy and Physiology of the Respiratory System

The human respiratory system, often taken for granted, is an intricate network designed for one fundamental purpose: sustaining life through gas exchange. From the initial intake of atmospheric air through the nasal passages and mouth, down the trachea, and into the branching bronchi and bronchioles, this system meticulously delivers oxygen to the alveoli. Here, in the microscopic air sacs of the lungs, oxygen is transferred to the bloodstream, simultaneously expelling carbon dioxide. This continuous, involuntary process is absolutely critical, making any disruption to its function an immediate and severe threat to an individual's operational capacity and, ultimately, their survival.

Critical Structures and Breathing Mechanics

Respiration, at its core, relies on a delicate interplay of skeletal and muscular components, each presenting a distinct vulnerability. The rigid cage formed by the ribs and sternum, while protective, can be compromised with focused impact, directly affecting lung function. Below this protective structure, the diaphragm, a crucial muscular sheet, orchestrates the rhythmic expansion and contraction essential for air exchange. Furthermore, the larynx and trachea, direct conduits for airflow, lie exposed and susceptible to immediate disruption.

The mechanical process of breathing is driven by precise pressure differentials, created through the synchronized action of these structures. Any compromise to the integrity of the thoracic cavity, the functionality of the diaphragm, or the patency of the airway directly impedes this critical function. Such disruption prevents proper lung inflation and gas exchange, rapidly leading to systemic oxygen deprivation. Understanding these interdependencies is paramount for effective tactical application.

Targeting these critical structures and disrupting their mechanics offers a direct pathway to rapid incapacitation. A precisely placed strike to the larynx or trachea can instantly occlude the airway, cutting off oxygen supply. Similarly, a concussive force applied to the rib cage can induce diaphragmatic spasm or fracture, leading to immediate respiratory distress. This approach aligns perfectly with C.E.T.A.'s objective of swift and decisive threat neutralization.

Methods of Respiratory Incapacitation

Respiratory incapacitation, within the C.E.T.A. framework, centers on immediately disrupting the body's essential oxygen exchange mechanisms. The objective is to halt or severely impede respiration, thereby rapidly depriving the brain and critical organs of oxygen. This swift cessation of function is achieved by targeting specific, vulnerable components of the respiratory system. Such methods bypass prolonged struggles, aiming for an involuntary and decisive outcome.

Direct trauma to the laryngeal or tracheal structures represents a primary method of immediate respiratory compromise. A forceful impact can induce laryngospasm, crush the cartilage, or even sever the trachea, leading to acute airway obstruction. Similarly, sustained compression of the thoracic cavity can prevent diaphragmatic excursion and lung inflation, effectively suffocating the individual. These actions exploit the mechanical fragility of the airway and the necessity of chest wall movement for breathing.

Beyond direct mechanical disruption, neural pathways controlling respiration offer another avenue for incapacitation. Specific pressure points or impacts to the cervical spine can disrupt the phrenic nerve's signal to the diaphragm, causing immediate respiratory arrest. Furthermore, any action that rapidly depletes the brain's oxygen supply, such as carotid artery compression, will quickly lead to unconsciousness and subsequent respiratory failure. Understanding these critical points allows for the precise application of force to achieve rapid incapacitation.

Physiological Consequences of Respiratory Failure

The abrupt cessation of oxygen delivery to the body's tissues marks the immediate consequence of respiratory failure. Without this vital gas, cellular respiration grinds to a halt, depriving cells of essential metabolic fuel. Within moments, blood oxygen saturation plummets, triggering a cascade of desperate physiological responses, including an accelerated heart rate and an involuntary struggle for breath. This acute hypoxia rapidly compromises motor control and cognitive function, rendering the target disoriented and increasingly incapacitated.

As oxygen deprivation persists, the brain, being the most metabolically demanding organ, suffers profound and irreversible damage. Cerebral hypoxia rapidly leads to loss of consciousness, typically within 30 to 60 seconds, followed by widespread neuronal death. Beyond the central nervous system, all other organ systems progressively fail as their cells are starved of oxygen, leading to complete systemic collapse. This rapid progression underscores why targeting the respiratory system offers one of the most direct and effective means of achieving swift physiological neutralization.

PFC Holifield, Platoon Leader, Basic Training Graduation, April 1987

AIT Class Leader, Ft. Sill, Oklahoma, April 1987

PFC Holifield, Platoon Leader, Ft, Sill, Oklahoma, 1987

Army Trainer Magazine, Hand-to-Hand Combat, August 1992

Army Trainer Magazine, Summer 1994

Black Belt Magazine, Defanging the Snake, April/ May 2021

Combat Warrior Magazine, Sikaron Karate, August 2007

Down and out

CHIEF COMBATIVES instructor at the Field Artillery Training Center Sgt. Leonard Holifield pins Staff Sgt. Carl Cantin, a drill sergeant from B Battery, 1st Battalion, 31st Field Artillery, during a hand-to-hand combat skill demonstration for soldiers-in-training. *See the story and more photos on Page 5A.*

CQC Basic Training Demonstration, Ft. Sill, Oklahoma

Fort Bragg Soldiers, Hand-to-Hand Combat Training, 1990

Fort Bragg CQC Training, 1992

COMPANY COMMANDERS WATCH AS TROOPS TRAIN FOR GULF WAR

Fort Bragg, NC, Hand-to-Hand Combat Training

Holifield with Korean Special Forces Chief CQC Instructor, Camp Casey, Korea 1993

Dr. Holifield with CSM Lesley Freeman Jr., with who he served on Camp Casey, Korea in 1993

Photos by Rick Jackson/Paraglide staff

Fighting machine!

bove: Holifield served as Chief Combatives Instructor, U.S. rmy from 1988-1997. Besides participating in the Comman- o Knock-Out Challenge 2, being held Friday, Holifield spent ils week training 18th Field Artillery Units in hand-to-hand- ombat. While stationed at Fort Bragg in 1990, he established program of hand-to-hand combat that was taught to units iroughout the post. Among his other successes, Holifield is redited with capturing one of the first Iraqi enemy prisoners f war during Operation Desert Storm.

ight: Prof. Leonard C. Holifield smashes 10 stacked bars of oncrete – 300 lbs. worth – during a demonstration of strength t the Commando Knock-Out Challenge 2 press conference t Fayetteville's Crown Coliseum.

"HOLIFIELD'S FIGHT FORFEITED"

Prof Leonard C Holifield, former Chief Combatives Instructor to the U S Army, was scheduled to fight in ***The USAvsWorld Commando Knockout Challenge 2*** on August 21, 1998 Howev his fights would be canceled due to his first opponent Colonel Mohamaad Jomaa, Head Instructor to the Lebanese Special Commando's and Presidential Elite Guard, forfeiting the bout due to problems securing his travel visa and the passing of his mother In a last ditch effort to secure an alternate opponent for Holifield, fight promoters flew in an opponent from Mexico, wh never showed up for the fight' "I understand Colonel Jomma's situation and my condolences go out his family, however, my opponent out of Mexico thought twice, and for his sake, may have made the right decision"

Holifield preparing for the USA vs The World CQC Championships. His Opponent, World Team Captain Mohammad Jomaa, Lebanon Special Forces and Chief Instructor to the Presidential Elite Guard, forfeited his fight with Holifield.

KORUS

MISSION

Sgt. Leonard C. Holifield, close quarters combat instructor, lashes out with a "lethal" strike to the head. This is just one of the many moves Holifield teaches to soldiers to defend themselves in a close combat situation.

Soldiers disable enemy with fists of steel

Story and Photos
by Nathan Maxfield

During Operation Desert Shield and Desert Storm, American soldiers, sailors, airmen and Marines demonstrated they could disable enemy equipment and engage enemy troops from miles away using state-of-the-art equipment.

But what happens when the enemy is closer? Much closer? What if the enemy is close enough to touch and your ammunition is gone?

Your training dictates that you take him out with your hands. Many people might not realize it, but one of the first enemy POW's captured during the Gulf War was taken by a U.S. soldier using hand-to-hand combat techniques.

Spec. Leonard C. Holifield, 5th Battalion, 8th Field Artillery, captured an Iraqi soldier who'd passed through the perimeter and was running toward the battalion tactical operations center yelling in Arabic.

Holifield didn't use a rifle, grenade or a bayonet. Holifield holds a 7th degree black belt in Sikaran Karate and easily disarmed and subdued the man.

"I think all soldiers should learn close-quarters combat," said Holifield, currently with 1st Battalion, 15th Field Artillery, Camp Casey. "I have taught infantry, special forces, rangers, recon units,

Martial artist shares expertise with Army

by John Valceanu
Staff writer

He moves slowly and deliberately, like a tiger stalking his prey. His six-foot, 225 body glides across a room with a surprising grace. Hard round muscles bulge outward, straining the shoulder seams of his uniform. It is readily apparent to anyone who sees him that many years of intense training were required to arrive at his present condition.

Holifield

This man is not only physically impressive however, he is also thoughtful and modest when talking about his past and his accomplishments.

Hegives the impression of someone who in posession of a very focused, tranquil mind. This effect is accomplished as much by what his soft voice leaves unsaid as by what is said.

Sgt. Leonard C. Holifield, Headquarters and Headquarters Battery, 5th Battalion, 8th Field Artillery Regiment, is a man who strove for personal excellence, achieved it, and now tries to help others achieve it through training them in combatives.

Holifield is a martial arts expert who freely imparts his expertise to others. His Sikaron Karate classes are offered free of charge to post servicemembers. He has helped train many units throughout the Army in the field of hand-to-hand combat. He is currently developing a Close Quarters Combat Instructor Certification program. Through this program he hopes to train commissioned and noncommissioned officers to be resident instructors for their units in close quarters combat.

the Cannoneer, June 8, 1995 5A

hand-to-hand combat

By Sgt. GINA BALTRUSCH
Field Artillery Training Center

"Sometimes the enemy is close enough to touch." Sgt. Leonard Holifield, the Field Artillery Training Center's chief combatives instructor, lives by those words. With 28 years of martial arts training and experience as qualifications, he has literally rewritten the way hand-to-hand combat training will be taught in the future by submitting recommended changes to the Army's training guidance, Field Manual 21-150 "Combatives."

Sgt. LEONARD HOLIFIELD

"My goal is to enhance the Army's hand-to-hand combat training, making it more battle-focused and safer for our soldiers," said Holifield, named 1994 instructor of the year by the World Martial Arts Hall of Fame.

"When I first joined the Army in 1987, the manual for hand-to-hand was almost 20 years old. It was updated in 1992, but the techniques are basically the same as the old one," he explained. "Some of those techniques can cause injury to the soldiers, and the training format just wasn't very realistic.

"Soldiers are performing combatives training in sand pits wearing tennis shoes. Not only will you never see a soldier wearing tennis shoes in a combat situation, but those shoes don't protect the ankle from injury during training. We're changing all that to a more battle-focused approach. Soldiers will wear protective helmets, web gear and combat boots — that's more like reality in actual combat."

The proposed improvements to the program include more than a change of clothing. Holifield modified many of the techniques, such as break-falls and defensive blocks, to enhance the safety aspects of training. He predicts the new program will render a 95 percent reduction of injuries caused by hand-to-hand combat training.

Holifield also added a few things to hand-to-hand training. Soldiers now learn ground fighting tactics, knife fighting (using flexible rubber knives for training) and choking techniques.

"The changes are very basic and easy to learn. The drill sergeants I've trained as assistant instructors seem to welcome the expertise I can give them. They seem real excited about the new program," said Holifield.

"The new methods are a lot safer than the old program...especially the break-falls. I've always wondered how to do a forward break-fall...now we have a training technique for it," said Staff Sgt. Carl Cantin, a drill sergeant at B Battery, 1st Battalion, 31st Field Artillery. "The biggest problem for us with aggressive physical training is the little injuries like twisted ankles and pulled muscles. Injuries delay soldiers' training. The new program really cuts down the injury risk and keeps our soldiers in training."

"I think the new changes have brought additional enthusiasm and confidence to the soldiers. The new skills are a plus in the training program," said Lt. Col. Michael Madden, 1st-31st FA commander, who recently observed the new hand-to-hand combat program.

Holifield anticipates leaving the FATC in the near future. He's recently been selected to head the unarmed self-defense program at the Army's Military Police School, Fort McClellan, Ala. However, he's making the most of the time he has left here. His days are filled with "train the trainer" classes, instructing hand-to-hand combat for new soldiers and developing a combatives training video with Fort Sill's TV Branch.

"That's the best I can give FATC before I go," Holifield said. "I dedicated myself to enhancing the Army's combatives programs when I was just a trainee at Fort Sill eight years ago."

"My personal goal for my career in the Army is to teach as many soldiers as I can how to fight and survive in a close combat environment," the world-certified combatives instructor explained. "This is a real-world approach to training that will give soldiers the skills and confidence they need to engage the enemy.

"With the military changing its focus from Cold War defense to more humanitarian missions, soldiers will be in close contact more often. Whether it's crowd control or actual face-to-face combat with an enemy, soldiers need to be able to subdue their opponents and survive."

Holifield training troops, Camp Casey, North Korea, 2ID

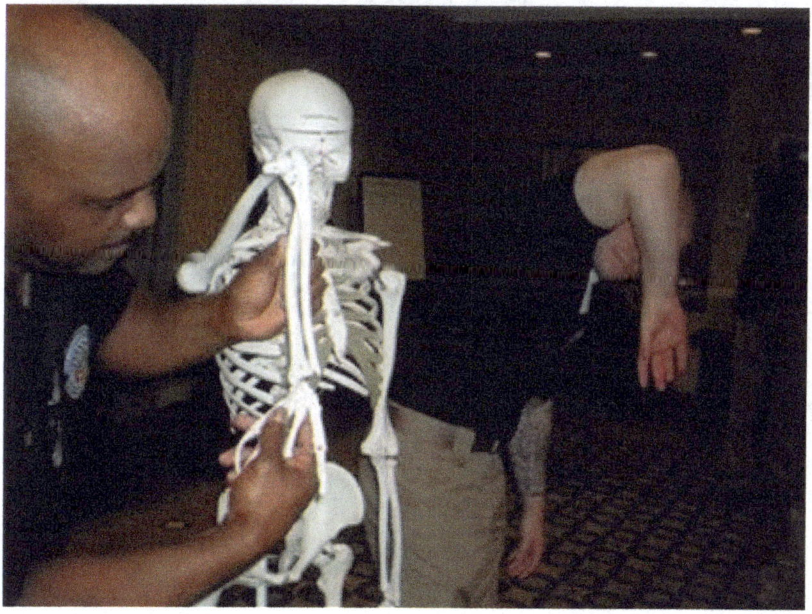

Holifield teaching C.E.T.A Skeletal System

U.S. ARMY CHIEF OF STAFF
GEN. GORDON R. SULLIVAN
CAMP CASEY KOREA - 1993

Holifield w/ Sergeant Major of the Army, Richard A. Kidd

Holifield with SMA Richard A. Kidd. Today's Army Combatives program is a result from this ground-breaking meeting.

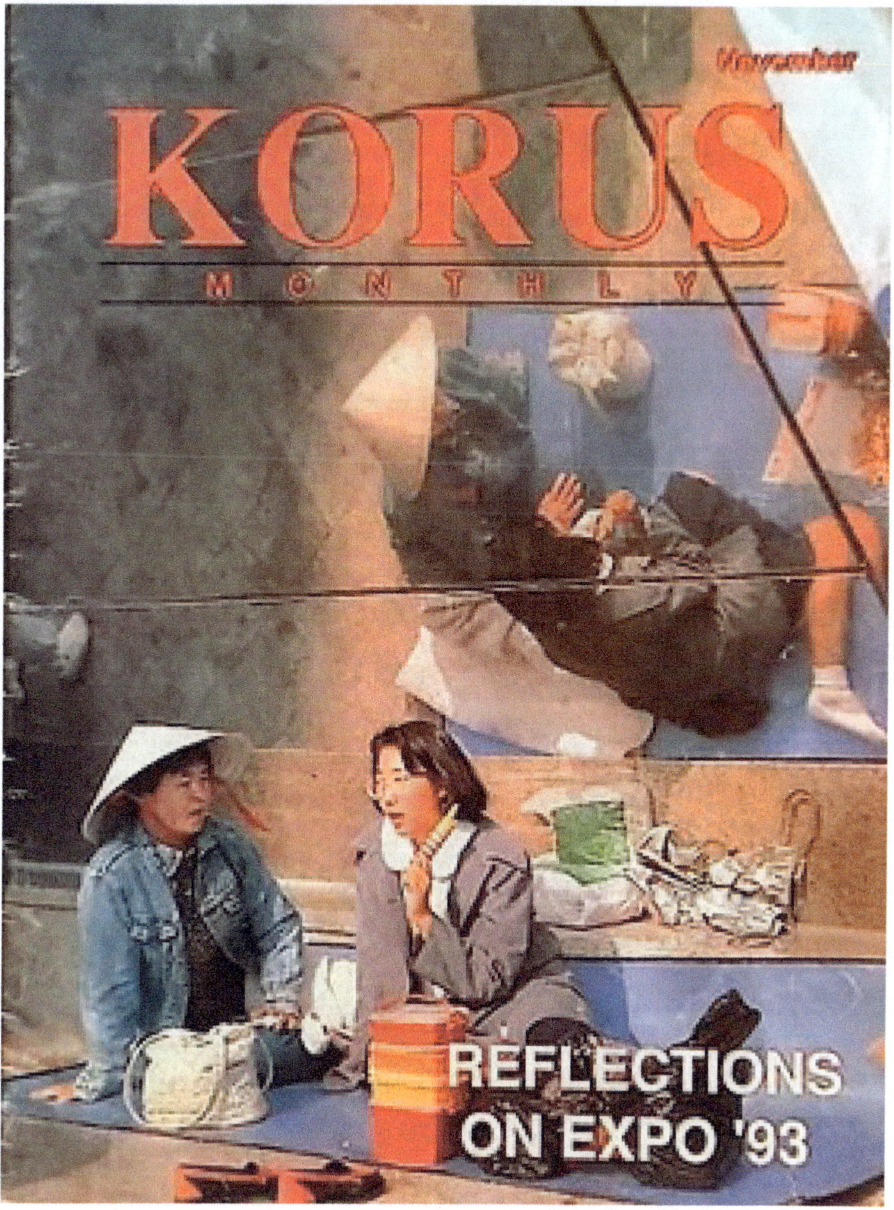

KORUS Korean Magazine, Fist of Steel article 1993.

Prof. Leonard Holifield Jug Tai Martial Artist Magazine, Greece

Holifield, Military Police Magazine, Ft. McClellan, AL.

Article "Are You Really Ready?"

Holifield, Military Police Officer, Ft. McClellan, AL

Executing missions alongside their counterpart gives each an appreciation of the other's role in fighting a battle. Lieutenants learn during Officer Basic that experienced noncommissioned officers are often quite capable of leading the platoon fight and that, as a team, the two can synergize their efforts and truly become a lethal battlefield force.

In other shared training, MP Officer Advanced Course students conduct a staff planning exercise to advise a battalion commander on how to best array forces to fight a rear battle. After completely staffing for a battalion operation, the captains are then "issued" four real second lieutenants and given the mission to fight the battle as a company commander. The idea of performing tactically before officers they have never met seems to bring out the very best efforts of all officers involved. Officers with the same projected assignments are grouped to create a team building exercise even before the officers reach their next assignment. Small Group Leaders from both courses prepare the exercise and coordinate after action review responsibilities. The end result is a more realistic exercise that produces more confident and more doctrinally reinforced battlefield leaders.

Captain Wayne P. Stilwell, was Chief, Military Police Warfighters' Complex, Directorate of Training, USAMPS, at the time this article was written.

UNARMED SELF-DEFENSE (ARE YOU REALLY READY?)

Staff Sergeant Leonard C. Holifield

The importance of unarmed self-defense (USD) training to the military police soldier cannot be over emphasized. At any given moment military police throughout the Army face the possibility of engaging in a physical confrontation with a suspect. This leads to the question: Are you really ready to deal with a violent suspect and are you confident in the USD skills that you may already possess?

The over-the-leg throw is an effective take-down technique

The Proper Mindset

Before USD skills can be used effectively, you must first develop the proper mindset which, in turn, will give you the confidence to engage with and effectively subdue a suspect. A positive mindset will help overcome the fear of intimidation caused by facing a person larger in physical size than you. This fear may also apply to hand-held weapons such as guns, knives, or other weapons. For most, the mere thought of looking down the barrel of a gun pointed at you is enough to cause you to "freeze." This could cause hesitation at the very moment you should react, which could put you and/or your partner in immediate danger. A positive mindset can be accomplished by constant training in realistic USD techniques geared toward real-world situations that military police soldiers are most likely to encounter. A positive mindset will allow you to react without hesitation and without fear of being injured. For the military police soldier, USD techniques are among the most valuable skills to possess. Sometimes these skills can mean the difference between life and death. When using effective USD techniques you must not only have confidence in your skills, but also possess a thorough knowledge of human anatomy, vital pressure points, and autonomic reflex, which is how the body might react to an applied technique. A good example of this is the application of a "joint control" technique. Manipulation of a joint–wrist, ankle, or elbow–could put a suspect in a position to be easily handcuffed.

58 MILITARY POLICE • PB 19-96-1

During his military career, Staff Sergeant Holifield (Above), served as Chief Defensive Tactics Instructor for the Military Police School, Ft. McClellan, AL; changing the course of unarmed self-defense for Military Police Officers army-wide.

Holifield Performing Knife Disarm, Giessen, West Germany, 1987

Holifield demonstrates flying scissor takedown, Ft. Bragg, NC. 1992

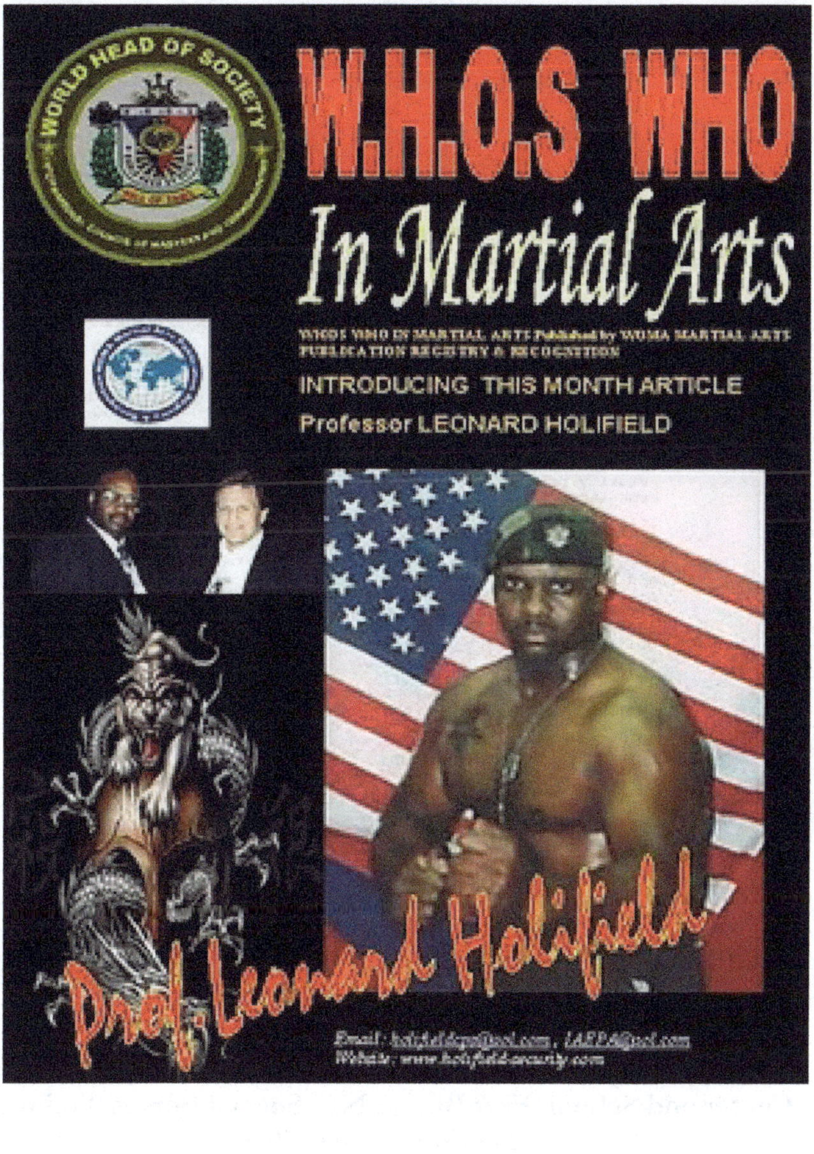
WORLD HEAD OF SOCIETY
W.H.O.S WHO
In Martial Arts
INTRODUCING THIS MONTH ARTICLE
Professor LEONARD HOLIFIELD
Prof. Leonard Holifield

DEPARTMENT OF THE ARMY
UNITED STATES ARMY JOHN F. KENNEDY SPECIAL WARFARE CENTER AND SCHOOL
FORT BRAGG, NORTH CAROLINA 28307-5000

REPLY TO ATTENTION OF: Commander, Headquarters & Headquarters Company,
5/3 Army Training Center, Army Post Office, NY. 09112

ATSU-SF 6 April 1988

SUBJECT: Special Forces Training Selection

1. Corporal Leonard C. Holifield, a member of your command has been selected to attend Special Forces Qualification Course 9-88. His tentative reporting date is 20 October 88, and his PCS orders should be received by your division levy section in about 4 weeks. Because of the nature of this training and the priority provided Special Forces Recruiting, requests for deferment or deletion will not be accepted.

2. I ask your support in assisting Corporal Holifield with this important career decision. Your assistance can be provided by insuring that he is in possession of an army rucksack (ALICE) and allowing him the time to complete the rigorous physical training (PT) program we have enclosed with his acceptance letter. This is critical to his ultimate success or failure and will be greatly appreciated by Corporal Holifield and his unit. A copy of the PT program has been enclosed for your review.

3. I sincerely appreciate your cooperation. Should any change occur in the applicants status, please notify the recruiting section at AUTOVON 239-5083 / 1818.

Encl

RICHARD W. POTTER, JR.
COLONEL, SPECIAL FORCES
ASSISTANT COMMANDANT

Acceptance Letter from the United States Army JFK Warfare Center and School, Fort Bragg, NC. Special Forces Training Selection. April 6, 1988

Top U.S. Military Combative Instructors of the 20th Century.

Dr. Holifield with assistant instructor James Turner, filming for Panther Productions Close-Quarter Combat video series.

Dr. Holifield with CQC Team, Camp Casey, Korea 1994

CQC TRAINING KOREA

6B the Cannoneer, September 3, 1987

Sports

THE OLD HAND IN THE FACE TRICK combined with tripping the enemy, kneeing him in the groin and breaking his knee with a palm thrust should work every time, provided his skills aren't equal to yours. Instructor PFC Leonard Holifield, in camoflage suit, gives Pvt. 1 Gary E. Baird a taste of his handy skill. Baird is with D Battery, 1st Battalion, 31st Field Artillery.

PFC Holifield Training Soldiers During Basic Training, Ft. Sill, OK. 1987

Hall of Fame Award for training U.S troops in hand-to-hand combat

Holifield, Camp Casey, Korea 1993

Resident tells of POW capture

By Ricardo Duran
Staff Writer

In the gritty reality of the battlefield, the phenomenon of the right man at the right time becomes commonplace. Call it fate or men and women rising to meet uncommon challenges.

For Leonard Holifield, 32, of Diamond Bar, capturing one of the first Iraqi prisoners of Operation Desert Storm was perhaps not as important as how he captured him.

"I had just come off of roving guard duty and it was around 1 a.m.," he said.

It was early morning of Feb. 12, nearly one month after the air war began, but not yet at the beginning of the ground war.

"I was back at the TOC [Tactical Operations Center] when we got a phone call from the perimeter guard that an unidentified individual was running toward us yelling in a non-English tongue."

The TOC was the nerve center of the headquarters of the 8th Field Artillery Regiment, which had been pounding frontline Iraqi positions across the Saudi-Kuwaiti border.

"We Heard an Iraqi. I guess he was an Iraqi, yelling in Arabic, and every once in a while he said 'Saddam, Saddam.' We locked and loaded, killed the lights and dropped to the floor," Holifield said.

Taking a pair of night vision goggles, Holifield crept out of the tent to subdue the Iraqi.

"I wasn't sure if he was alone, if it was a setup...I had no idea what he was doing," he said. "I felt scared, you know, fear of the unknown. All I knew was that I had a responsibility — I had to take charge."

Though nursing a broken wrist, the seventh degree black belt in Sikaron — an eclectic Filipino martial art — rushed the Iraqi, threw him to the ground in one swift judo-like motion and turned him onto his stomach.

"I secured his neck, arm and wrist and he settled down and relaxed," Holifield said.

Holifield is a hand-to-hand combat instructor at Ft. Bragg, North Carolina.

"I felt confident in my abilities," he added, although fearing the possible dangers involved. "My life and my family flashed before me. I felt a kaleidoscope of feelings."

The soldier turned out to be a defector from an Iraqi air defense unit.

"We gave him food, water and some cigarettes.... All I know is that his name was Ali," Holifield said. "He was cold, hungry and frightened of being killed."

As a specialist in hand-to-hand combat, Holifield knows it takes both physical as well as psychological training to fight effectively.

"The skill to kill must be matched by the will to kill," he said.

Still, Holifield is a soft-spoken humble man, ascribing his humility to the philosophical aspects of the martial arts.

"It develops in you a base of self control. You must be confident, but must also have humbleness," he said. "Humbleness in knowing what you can do and what you cannot. You must never abuse the (martial) arts."

In a business so filled with senseless death, perhaps Holifield was the right man at the right time — to the benefit of one hungry, frightened Iraqi soldier.

Staff photo by Ricardo Duran

Hero's welcome: Army Specialist Leonard Holifield receives a standing ovation following presentations by representatives of the city, Rep. David Dreier's and Assemblyman Paul Horcher's offices at the May 7 City Council meeting.

Homecoming welcome for Holifield, following the Gulf War. Honored by the Mayors office for capturing the first Iraqi Prisoner of War during Operation Desert Storm 1990.

POW CERTIFICATE OF RECOGNITION FROM CALIFORNIA STATE LEGISLATURE

FIRST POW CAPTURE, GULF WAR

COL LEWIS AND CSM STEWART

Soldiers Magazine Article "Hand-to-Hand Combat" Dec 1995

Supreme Grandmaster Holifield, 10th Dan, Soke
Sikaron Karate Federation

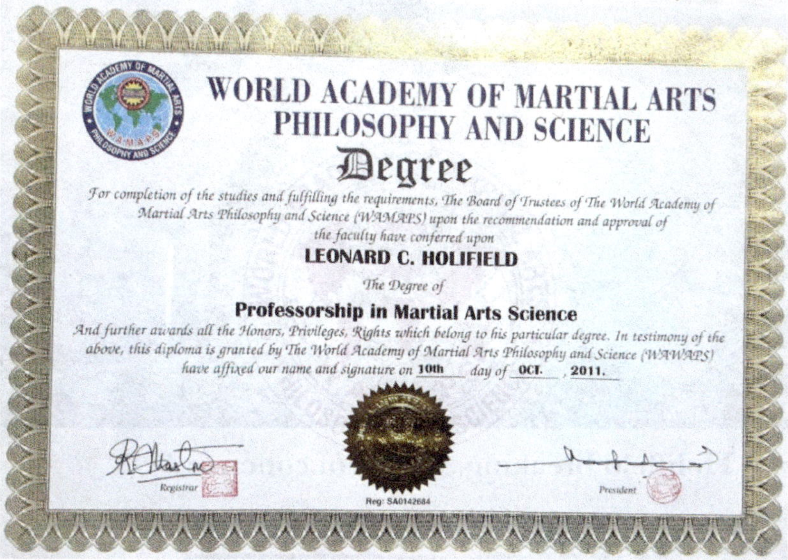
WORLD ACADEMY OF MARTIAL ARTS
PHILOSOPHY AND SCIENCE
Degree
For completion of the studies and fulfilling the requirements, The Board of Trustees of The World Academy of Martial Arts Philosophy and Science (WAMAPS) upon the recommendation and approval of the faculty have conferred upon
LEONARD C. HOLIFIELD
The Degree of
Professorship in Martial Arts Science
And further awards all the Honors, Privileges, Rights which belong to his particular degree. In testimony of the above, this diploma is granted by The World Academy of Martial Arts Philosophy and Science (WAWAPS) have affixed our name and signature on 10th day of OCT. , 2011.
Registrar
President
Reg: SA0142684

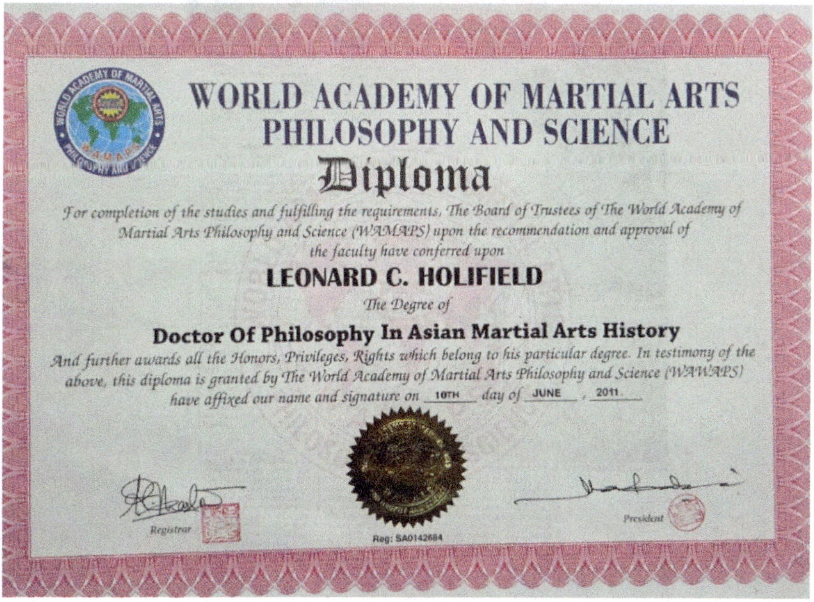
WORLD ACADEMY OF MARTIAL ARTS
PHILOSOPHY AND SCIENCE
Diploma
For completion of the studies and fulfilling the requirements, The Board of Trustees of The World Academy of Martial Arts Philosophy and Science (WAMAPS) upon the recommendation and approval of the faculty have conferred upon
LEONARD C. HOLIFIELD
The Degree of
Doctor Of Philosophy In Asian Martial Arts History
And further awards all the Honors, Privileges, Rights which belong to his particular degree. In testimony of the above, this diploma is granted by The World Academy of Martial Arts Philosophy and Science (WAWAPS) have affixed our name and signature on 10TH day of JUNE , 2011.
Registrar
President
Reg: SA0142684

Holifield breaking 300 lbs of concrete 1998

SOLDIERS

OCTOBER 1990

Holifield: Training troops in hand-to-hand.

What do you do in combat if your weapon jams or gets knocked out of reach? Just ask Spec. Leonard C. Holifield, chief combatives instructor for Headquarters Battery, XVIII Airborne Corps Field Artillery at Fort Bragg, N.C. Holifield, a seventh-degree black belt in Philippine Sikaron karate, has used his 23 years of experience in martial arts to train troops in hand-to-hand combat during tours at Fort Sill, Okla., and in Giessen, West Germany. Since arriving at Fort Bragg in March, he's been coordinating with unit commanders to train soldiers there as well.

Pointing out that hand-to-hand combat is a last resort rather than a replacement for a soldier's weapon, Holifield said "more countries are now training in hand-to-hand than the U.S." The Army could be at a disadvantage in any future close-quarters combat, he said, because of its relative lack of martial arts training.

As southeastern regional director of the Armed Forces Martial Arts Association and the North American Combat Academy, Holifield is working to increase military hand-to-hand skills by supporting the creation of "a cohesive and organized, Army-wide combatives training program."

Holifield said military martial arts training has a number of positive results, including "high morale, teamwork, unit cohesion and improved physical fitness." However, he says the skills he teaches prepare soldiers for daily life too.

"Martial arts training instills mental discipline," Holifield pointed out, "something that is vital both on and off the battlefield." — *Scott Rogers, Fort Bragg PAO*

Thirty years after taking an oath to serve her country, the highest ranking black enlisted female, **CSM Jannette Moorer**, is closing the book on a highly decorated career.

Born in Walterboro, S.C. in 1942, Moorer joined the Army a year after graduating from high school. Although her parents were against her decision, she insisted it was what she wanted to do.

"My father thought it was crazy. He refused to sign for me, but my mother eventually convinced him that if he didn't, I would wait until I was old enough," she recalled.

Starting her career at a time when opportunities for women and minorities were limited, she decided early not to let her own reaction to discrimination get in the way of success. "I knew it was there, but I was never confronted personally. I knew that as a black female, it was going to be hard," she said.

On Nov. 1, 1980, Moorer became the first Regular Army black female to be promoted to Command Sergeant Major.

Since then, she has served two tours in Korea, one as a post commander, and tours at Fort Jackson, S.C., including a year as CSM of an infantry training battalion.

"She's a super human being who's straight forward and honest," says CSM Arthur Hart, commandant of the Fort Jackson Drill Sergeant School. "If you want to get beat up, mess with any of her soldiers. Both male and female soldiers can learn a lot from her." — *Sheila Safey, Fort Jackson PAO*

The Army Community Services building at Fort Lewis, Wash., has a new addition. Normally, such an announcement wouldn't receive much attention. But it has a special importance for **Spec. Mariusz Kowalewski**. The newly sworn-in U.S. citizen is the designer of record.

For Kowalewski, a member of HHC 864th Engineer Bat-

42

SOLDIERS

CHAPTER 5

THE BRAIN

THE SEAT OF CONSCIOUSNESS

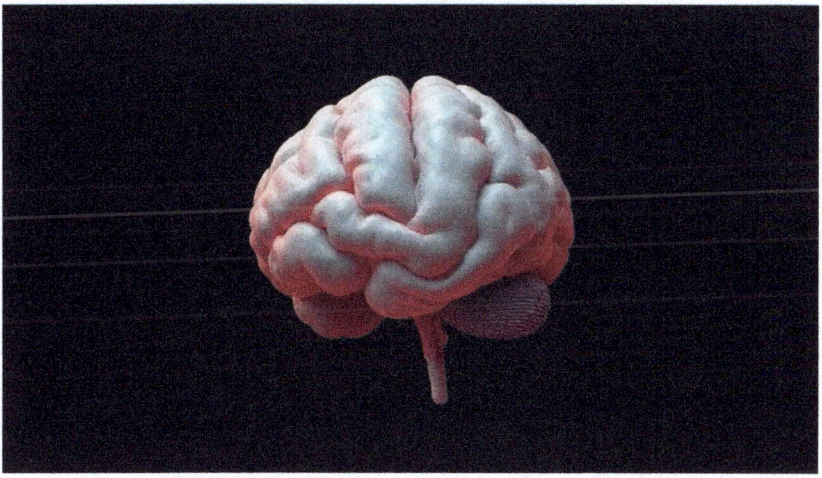

Neuroanatomy and Brain Function

The human brain functions as the ultimate command and control center, orchestrating every physiological process and conscious thought. Its intricate structure, encompassing the cerebrum, cerebellum, and the critically important brainstem, forms a complex network. This master organ dictates everything from voluntary movement to involuntary respiration and heartbeat, making its integrity non-negotiable for survival. Within the C.E.T.A. framework, understanding its foundational neuroanatomy is paramount, as disrupting this central hub inevitably cascades into systemic failure.

The brainstem, specifically, serves as the non-negotiable nexus for all life-sustaining functions, housing the centers vital for breathing, circulation, and consciousness. Even a momentary interruption to its integrity can lead to immediate and irreversible cessation of these critical processes. This inherent vulnerability underscores why precise targeting of the cranium and upper cervical spine can yield rapid incapacitation. Mastery of these anatomical specifics is not merely academic; it is the cornerstone of calculated destruction.

Brain Vulnerabilities and Impact Points

While encased in the formidable skull, the brain possesses critical points of vulnerability that, when targeted, can lead to immediate and profound incapacitation. The brain itself, a soft, gelatinous organ, is suspended within cerebrospinal fluid, making it highly susceptible to sudden rotational or concussive forces. Understanding these specific anatomical weaknesses is paramount for effective close-engagement targeting. It's not merely about applying brute force, but rather about the precise application of energy where the body's natural defenses are inherently compromised.

Key impact points include the temples, where the skull's bone structure is notably thinner, allowing for more direct force transmission to the frontal and temporal lobes. Strikes directed to the jaw or chin can induce rapid head acceleration, causing the brain to violently collide with the interior of the cranium, often resulting in a severe concussive state or immediate loss of consciousness. Furthermore, direct trauma to the mastoid process or the delicate base of the skull can transmit shockwaves directly to the brainstem, disrupting vital autonomic functions and potentially leading to instantaneous neurological shutdown. These specific areas offer the most efficient pathways for rapid incapacitation.

Techniques for Brain Disruption

Disrupting brain function primarily involves the rapid transfer of kinetic energy to the cranial vault. This energy, when delivered with sufficient force, induces traumatic brain injury, ranging from concussion to more severe structural damage. The objective is to overwhelm the brain's protective mechanisms and impair its operational capacity. Strikes must be precise, targeting areas that maximize intracranial pressure or direct neural shock. Understanding vector mechanics is crucial for achieving immediate incapacitation.

Beyond direct percussive force, disruption can also stem from compromising cerebral blood flow. Techniques targeting the major arteries supplying the brain, such as the carotid or vertebral arteries, can induce rapid hypoxia. Even a momentary cessation of blood flow can cause immediate disorientation and loss of consciousness. Furthermore, repetitive or sustained sub-concussive impacts can cumulatively degrade neurological function, though this is less suited for immediate threat neutralization. The goal remains swift and absolute functional compromise.

Neurological Effects of Trauma

Direct trauma to the cranium, particularly impacting the brain, elicits immediate and profound neurological responses. The sudden concussive force disrupts the delicate neural pathways, leading to an instantaneous cascade of electrochemical disarray. This initial shock often manifests as a loss of consciousness, disorientation, or severe motor dysfunction, effectively compromising the target's ability to respond or resist.

Beyond the immediate impact, brain trauma can initiate a complex series of secondary neurological events. Edema, hemorrhage, and intracranial pressure rapidly develop, further compressing vital brain structures and impeding normal neuronal function. These internal processes disrupt the brain's regulatory control over essential bodily systems, extending the incapacitating effects far beyond the point of initial contact.

The resulting neurological deficit can range from transient impairment to permanent vegetative states or fatality, depending on the severity and location of the trauma. Understanding these cascading effects is crucial for tactical applications, as it underscores the importance of precision in targeting to achieve rapid and irreversible neurological compromise. The goal is not merely to inflict pain, but to systematically dismantle the central control mechanisms governing conscious action and physiological stability.

CHAPTER 6

THE SKELETAL SYSTEM

FRAMEWORK AND SUPPORT

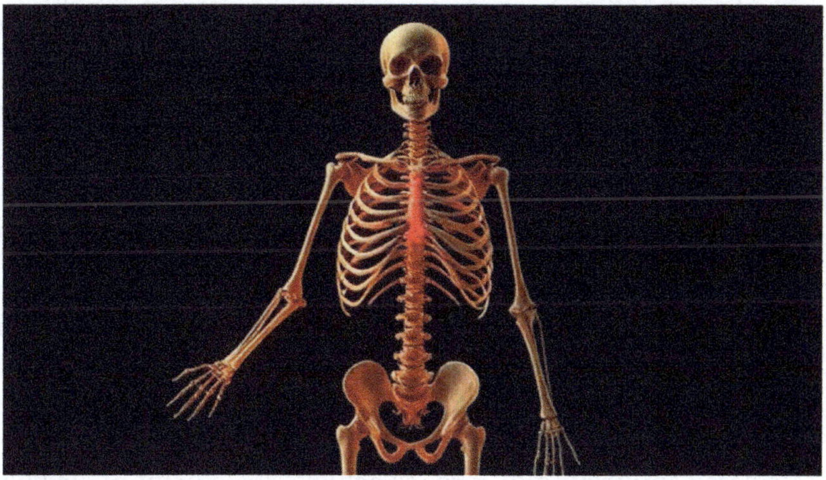

Skeletal Anatomy and Biomechanics

The human skeletal system serves as the foundational framework, providing essential support, protection for vital organs, and anchor points for muscular movement. Comprised of bones, cartilage, and ligaments, this intricate structure enables mobility and withstands significant stress. However, despite its apparent robustness, the skeleton possesses inherent biomechanical weaknesses that, when precisely understood, become critical targets for incapacitation. Identifying these structural vulnerabilities is paramount in the strategic application of force within close-engagement scenarios.

Understanding the biomechanics of the skeleton goes beyond mere identification of bones; it necessitates a comprehensive grasp of how forces are distributed and absorbed throughout the system. Joints, as pivotal points of articulation, represent common loci for disruptive force application, while variations in bone density and shape dictate their individual resistance thresholds. Exploiting these mechanical principles allows for the calculated generation of leverage, leading to structural failure and the rapid neutralization of an adversary. This analytical approach transforms the skeletal system from a protective shield into a detailed map of exploitable weaknesses.

Fracture Points and Structural Weaknesses

The human skeletal system, while a marvel of engineering, possesses distinct structural weaknesses that, when exploited, can lead to immediate and profound incapacitation. These vulnerabilities are not random but stem from biomechanical realities: areas of concentrated stress, reduced bone density, or minimal soft tissue protection. Identifying these inherent fracture points moves beyond mere brute force, allowing for the calculated application of force to achieve rapid orthopedic failure. Such an understanding is fundamental to the C.E.T.A. methodology, focusing on precision over power.

Consider, for instance, the long bones, which, despite their strength, exhibit specific weak zones. The shaft of the tibia, particularly susceptible to direct impact due to its superficial nature, readily yields to focused trauma. The clavicle, a slender and exposed bone, fractures with relative ease, disrupting shoulder girdle function and causing intense pain. Furthermore, the ribs, while offering protection, are vulnerable to specific angles of compression or strike, posing a direct threat to the thoracic organs once compromised.

Targeting the Skeleton for Incapacitation

Building upon our understanding of the skeleton's inherent vulnerabilities, effective incapacitation often hinges on disrupting its structural integrity. The goal is not merely to inflict pain, but to compromise the body's foundational support, rendering an opponent unable to maintain posture or execute motor functions. This deliberate targeting exploits the bone's role as a lever and anchor for muscle, turning strength into weakness. By focusing on specific skeletal points, an operator can rapidly disable a threat, moving beyond superficial injury.

Key targets for skeletal incapacitation include long bones, major joints, and load-bearing structures. For instance, a well-placed strike to the tibia or fibula can induce a compound fracture, immediately collapsing an opponent's stance. Similarly, dislocating a major joint like the shoulder or knee severs the functional connection between segments, making movement impossible. The clavicle, despite its relatively small size, is also a highly effective target, as its fracture severely impairs arm and shoulder mobility.

The immediate physiological response to severe skeletal trauma involves intense pain, shock, and a profound loss of mobility. This rapid onset of incapacitation is the tactical objective, preventing further aggression or resistance. Beyond the immediate physical collapse, such damage initiates a cascade of systemic responses that further diminish an individual's capacity to fight. Understanding these direct and secondary effects is crucial for predicting and ensuring the desired outcome in a close-engagement scenario.

Consequences of Skeletal Trauma

Skeletal trauma, particularly fractures or dislocations, immediately introduces profound pain and structural instability. This direct assault on the body's framework instantly compromises an individual's ability to maintain posture, execute motor functions, or apply force effectively. Even a partial fracture can induce significant localized pain, diverting critical focus and impairing coordinated movement. The resulting incapacitation is often immediate, rendering the target unable to continue engagement.

Beyond the initial mechanical failure, severe skeletal trauma frequently precipitates a cascade of secondary physiological complications. Sharp bone fragments can lacerate adjacent vasculature, leading to rapid internal hemorrhage and hypovolemic shock. Furthermore, nerve impingement or transection by displaced bone can result in immediate loss of sensation and motor control distal to the injury site. These concurrent injuries amplify the body's systemic shock response, accelerating the target's functional collapse.

The strategic application of force to critical skeletal points aims not merely for pain, but for definitive structural compromise. A shattered joint or a vertebral fracture, for instance, ensures not only immediate physical collapse but also prolonged, debilitating incapacitation. Such injuries effectively remove the individual from any threat matrix, fulfilling the C.E.T.A. objective of rapid and irreversible neutralization. This systematic destruction of the body's foundation is paramount in high-stakes engagements.

Complications of broken bones include healing issues like nonunion (failure to heal) or malunion (healing in a bad position), soft tissue damage from the bone's sharp edges, nerve damage, blood vessel damage leading to bleeding or lack of blood flow, fat or pulmonary embolism, and infection. Severe breaks can also result in long-term problems such as post-traumatic arthritis, joint instability, chronic stiffness, and, in rare cases, the need for amputation due to complications like compartment syndrome.

Healing and Bone Issues

- **Delayed Union:** The bone takes longer than expected to heal. bone heals but in a misaligned position.
- **Nonunion:** The bone fails to heal completely at all.
- **Osteonecrosis:** A lack of blood flow causes a portion of the bone to die.
- **Post-traumatic arthritis:** Fractures extending into joints can damage cartilage, leading to arthritis and joint problems.
- **Leg Length Discrepancy:** In some cases, a fracture can result in unequal leg lengths.

Soft Tissue & Nerve Complications

- **Soft tissue damage:**The broken bone can tear surrounding muscles, nerves, and blood vessels.
- **Nerve damage:** A bone fracture or displaced bone can stretch or crush nerves, leading to numbness or paralysis.
- **Blood Vessel Damage:** Severe fractures, particularly in the femur or pelvis, can cause significant blood loss or restrict blood flow to the limb.
- **Compartment Syndrome:** Pressure builds up in the muscles, cutting off blood supply, which can cause permanent nerve and muscle damage and may require amputation.

Systemic & Other Complications

- **Infection:** Bacteria can enter during the injury (especially open fractures) or surgery, leading to bone infections like osteomyelitis.
- **Pulmonary Embolism:** A blood clot (often from a DVT) can travel to the lungs, a severe complication of hip or pelvic fractures.
- **Fat Embolism:** Fat released from the bone marrow of long bones can travel to the lungs, causing respiratory distress.

Long-Term Issues

- **Joint Instability:** Fractures can destabilize nearby joints, increasing the risk of osteoarthritis.
- **Stiffness and Impaired Range of Motion:** Scarring and prolonged immobilization can lead to stiffness, especially in joints like the knee, elbow, and shoulder.

CHAPTER 7

INTERNAL ORGANS

VITAL FUNCTIONS AND VULNERABILITIES

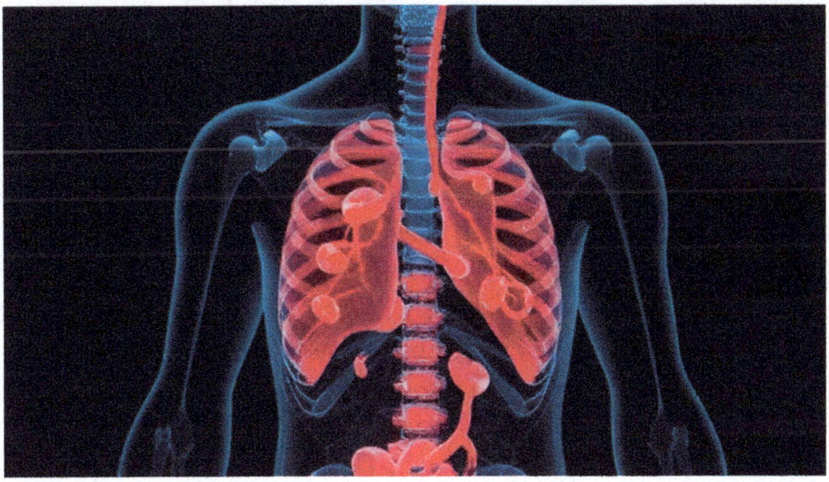

Overview of Major Internal Organs and Their Functions

Within the intricate architecture of the human torso lies a complex array of internal organs, each performing vital functions essential for survival. While often shielded by skeletal structures, these visceral components represent critical points of failure when subjected to targeted trauma. A comprehensive understanding of their basic physiological roles is paramount for any operative seeking to effectively neutralize a threat. Disrupting these fundamental processes directly compromises the body's ability to sustain life, leading to rapid incapacitation.

The liver, a massive and highly vascular organ, manages detoxification and metabolism, while the spleen plays a crucial role in blood filtration and immune response. Below the diaphragm, the stomach and intestines are central to digestion and nutrient absorption, processes critical for energy and cellular repair. Damage to any of these soft, delicate structures can result in catastrophic internal bleeding, sepsis, or systemic shock. Such injuries rapidly degrade an individual's combat effectiveness, transforming a capable adversary into a medical emergency.

Vascular and Visceral Vulnerabilities

The human body, a marvel of interconnected systems, paradoxically harbors numerous points of profound vulnerability, particularly within its vascular and visceral networks. These systems, while essential for life, are often surprisingly exposed or fragile when subjected to focused trauma. Understanding these inherent weaknesses is paramount for anyone seeking to effectively neutralize a threat, as their compromise can lead to immediate systemic collapse or profound, irreversible damage. The meticulous study of these vulnerabilities forms the bedrock of close-engagement target acquisition.

The vascular system, a pressurized network of blood delivery, presents immediate and catastrophic targets. Major arteries like the carotid in the neck, the femoral in the groin, or the brachial in the upper arm, when severed or ruptured, result in rapid exsanguination and precipitous drops in blood pressure, leading to profound shock and unconsciousness within seconds. Even significant venous structures, though under lower pressure, can cause fatal hemorrhage if compromised deeply enough. Targeting these pathways ensures a swift and decisive cessation of physiological function.

Beyond the circulatory network, the visceral organs themselves offer distinct points of attack, despite varying degrees of natural protection. Organs such as the liver, spleen, and kidneys, though partially shielded by the ribcage or abdominal musculature, are highly vascularized and prone to catastrophic internal bleeding upon blunt force trauma or penetration. Damage to the intestines can lead to peritonitis and sepsis, while even non-fatal but debilitating strikes to organs like the stomach or bladder inflict agonizing pain and incapacitation. Each visceral target, when properly engaged, contributes to the rapid and decisive neutralization of an assailant.

Tactical Targeting of Internal Organ Systems

While the skeletal and central nervous systems present evident targets, the internal organ systems, though often protected, offer direct pathways to rapid incapacitation when precisely engaged. Their disruption directly compromises the body's essential life support functions, leading to systemic collapse. Effective tactical targeting, therefore, necessitates a profound understanding of anatomical landmarks and the likely vectors of force required to bypass natural defenses. This precision allows for the calculated destruction vital for achieving swift neutralization in high-stakes encounters. Such strikes are not about brute force, but about applied anatomical knowledge.

Executing a successful strike against an internal organ demands a precise application of kinetic energy, often through focused blunt trauma or directed compression. Targets like the spleen or liver, while somewhat shielded, remain vulnerable to severe impact due to their friable nature and rich vascularity. Similarly, the kidneys, though posterior, can be compromised through forceful strikes to the flanks, leading to significant hemorrhage. Understanding the specific physiological response to such trauma, from immediate shock to progressive systemic failure, is paramount for predicting and ensuring the desired incapacitation effect. This deliberate targeting transforms a simple strike into a strategic act of disablement.

Physiological Impact of Internal Organ Damage

When internal organs sustain trauma, the immediate physiological response is often severe and systemic. Highly vascularized organs like the liver or spleen can bleed profusely, leading to rapid internal hemorrhage. This acute blood loss quickly causes hypovolemic shock, drastically lowering blood pressure and depriving critical tissues of oxygen. The body's homeostatic balance is immediately compromised, initiating a cascade of system failures.

Beyond bleeding, the specific function of the damaged organ dictates further physiological decline. A punctured lung, for instance, causes pneumothorax, severely restricting respiration and oxygen absorption. Trauma to the kidneys can lead to acute renal failure, allowing metabolic waste to rapidly accumulate within the system. Intestinal damage risks peritonitis, a severe infection that often progresses to septic shock; each injury directly compromises a vital life support system, with consequences escalating rapidly.

The true danger of internal organ damage often lies in its initially subtle presentation, concealing severe trauma until critical thresholds are passed. Once a vital organ's functional capacity diminishes significantly, the body's compensatory mechanisms quickly fail. This rapid deterioration highlights the efficacy of targeting these systems for swift incapacitation, as the physiological aftermath is inherently catastrophic. The body simply cannot recover from such fundamental systemic disruption.

Internal organ damage causes systemic physiological impacts, including hemorrhage leading to shock, impaired organ function, infection, and pain, often manifesting with symptoms like abdominal swelling, bruising, or difficulty breathing. Damage to one organ can disrupt its surrounding systems, leading to a cascade of complications, such as respiratory or circulatory problems, and can result in chronic pain or disability. Treatment involves addressing the internal bleeding and organ dysfunction, and long-term outcomes depend on the injury's severity, the individual's health, and the effectiveness of medical care.

Immediate Physiological Impacts

- **Hemorrhage and Shock:** Damaged organs can bleed internally, reducing blood volume and leading to shock due to decreased blood flow to vital organs.
- **Organ Dysfunction:** Damaged organs cannot perform their functions effectively, potentially leading to acute organ failure.
- **Swelling and Compartment Syndrome:**Swelling around the damaged organ can increase pressure, cutting off blood flow to the organ and further damaging its tissues.
- **Infection:**Compromised organs, especially the digestive tract, can become entry points for bacteria, leading to systemic infections.

Disruption of Body Systems

- **Circulatory System:**Damage to the heart can cause issues with blood clots, heart attacks, or strokes.
- **Respiratory System:** Injury to the lungs or surrounding organs can hinder breathing, potentially causing respiratory failure.

Signs and Symptoms

- **Pain and Tenderness:** Pain or tenderness in the affected area is a common symptom.
- **Swelling and Bruising:** Visible bruising or abdominal bloating can indicate internal injury.
- **Respiratory Distress:** Difficulty breathing, shortness of breath, or rapid breathing can signal damage to the respiratory system.
- **Abdominal Symptoms:** Nausea, vomiting, abdominal rigidity, or distension.
- **Circulatory Issues:** Dizziness, fainting, lightheadedness, rapid heartbeat, or an irregular pulse.
- **Skin Changes:** Jaundice (yellowing of the skin and eyes) can indicate liver damage.

Long-Term Complications

Chronic Pain and Disability: Survivors may experience ongoing pain and physical limitations.

Psychological Effects: Anxiety, depression, and stress can arise from the injury and its impact on daily life.

Reduced Quality of Life: The cumulative effects of organ damage can significantly impact a person's well-being and daily functioning.

Management and Prognosis

Medical Intervention: Treatment may involve surgery to control bleeding or repair organs, blood transfusions, intravenous fluids, and medication.

- **Rehabilitation:** Ongoing care and lifestyle adjustments are often necessary.

- **Prognosis:** The outcome depends on the severity of the organ damage, the speed of diagnosis and treatment, and the individual's overall health.

CHAPTER 8

INTEGRATED SYSTEMS ANALYSIS IN COMBAT

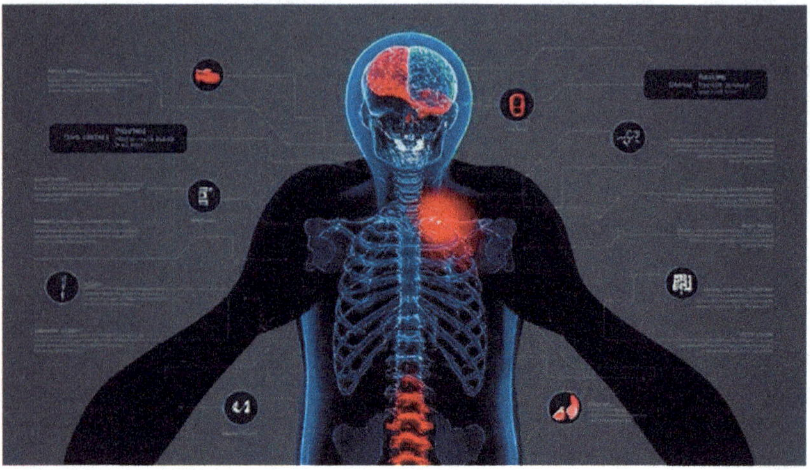

Synergistic Effects of Multiple System Trauma

While the study of individual system vulnerabilities provides a foundational understanding, true proficiency in close-engagement threat neutralization demands an appreciation for synergistic effects. When multiple life support systems endure trauma concurrently, the resulting physiological collapse accelerates exponentially. This compounded damage disrupts the body's compensatory mechanisms, leading to a far more rapid and profound incapacitation than any single isolated strike could achieve. Understanding this principle is crucial for effective tactical application.

Consider, for instance, the combined impact of a significant respiratory compromise coupled with severe circulatory disruption. The immediate lack of oxygen delivery, exacerbated by insufficient blood flow to vital organs, rapidly starves the brain and heart. Such an attack bypasses the body's natural resilience, pushing the individual into an irreversible state of systemic failure within seconds. This principle underscores the C.E.T.A. methodology's emphasis on comprehensive targeting.

Therefore, a strategic approach to threat neutralization prioritizes not merely damaging a single critical system, but orchestrating a cascade of failures. Targeting the central nervous system while simultaneously impeding the respiratory or circulatory pathways creates a devastating and often immediate incapacitation. This calculated destruction maximizes efficiency, ensuring that the target's ability to resist or counter is irrevocably compromised. Mastery of these synergistic principles is what elevates tactical engagement from mere striking to precise, calculated destruction.

Assessing Threat Levels and Target Priorities

In any close-engagement scenario, the initial and most critical phase is the accurate assessment of the threat. This is not merely about identifying an aggressor, but rather quantifying their immediate danger and potential for harm. A swift, precise evaluation allows for the optimal application of C.E.T.A. principles, ensuring that response is proportionate yet decisive. Misjudgment here can lead to either over-application of force or, more dangerously, insufficient action.

Once the threat level is established, the tactical operator must immediately pivot to prioritizing targets. This involves a rapid analysis of the assailant's posture, movement, and potential for counter-action, directly informing which of the body's critical life support systems offers the most immediate and devastating point of attack. The goal is always to achieve the quickest possible cessation of the threat. Effective prioritization dictates the most efficient pathway to incapacitation or neutralization.

Several factors influence this target prioritization beyond the initial threat assessment. Environmental considerations, such as confined spaces or obstacles, might limit access to certain targets, necessitating an adaptive approach. Furthermore, the aggressor's physical attributes, including their size, musculature, or even their current state of awareness, can shift the optimal point of impact. Ultimately, the chosen target must align with the tactical objective: to disrupt the threat's ability to continue the engagement.

Case Studies in Tactical Engagement

The transition from theoretical understanding to practical application demands rigorous examination of specific engagement scenarios. These case studies serve as critical analytical tools, illuminating how C.E.T.A. principles are applied in dynamic, high-stakes environments. Through the meticulous dissection of both historical and hypothetical situations, we gain invaluable insight into the calculated disruption of life support systems. Each scenario offers a distinct perspective on threat neutralization, emphasizing the precise execution of targeting protocols.

These detailed analyses will further demonstrate the intricate process of rapid threat assessment and the subsequent prioritization of critical physiological targets. We will observe the tangible manifestation of synergistic trauma, illustrating how concurrent compromise of multiple bodily systems accelerates incapacitation. Such examinations reinforce the absolute necessity of swift vulnerability identification and exploitation in combat. Ultimately, understanding these engagements, refines tactical responses and deepens comprehension of C.E.T.A.'s unparalleled efficacy.

Adapting C.E.T.A. to Evolving Threats

The operational landscape is rarely static; indeed, it constantly shifts, presenting new challenges to even the most refined tactical systems. While the human body's fundamental vulnerabilities remain constant, the methods by which adversaries engage, protect themselves, or utilize novel technologies can profoundly alter threat dynamics. Therefore, the sustained efficacy of C.E.T.A. relies not just on its foundational principles but equally on its inherent capacity for adaptation. This ensures the system remains optimally effective for rapid threat neutralization, regardless of evolving variables.

Adaptation within the C.E.T.A. framework does not necessitate a fundamental overhaul of its core tenets—the targeting of critical life support systems remains paramount. Instead, it involves a continuous refinement of application, incorporating new data from combat analysis, physiological research, and material science advancements. For instance, the proliferation of advanced ballistic protection or the emergence of novel combat methodologies demands a fluid re-evaluation of entry points and force vectors. This ongoing analytical process ensures C.E.T.A. principles translate into effective countermeasures against any contemporary threat.

At CETA's core is a lethal combination of martial art techniques dissected from Judo, Jujitsu, Aikido, Hapkido, Combat Karate, Shotokan Karate, Ninjitsu, Hwangdo, Advanced Knife-Fighting, Boxing, Grappling, Krav-Maga, and Dim-Mak. That being said, CETA can evolve seamlessly into effective fighting techniques that suit most any violent physical attack. Adapting self-defense to evolving threats requires continuous development of situational awareness, de-escalation skills, and flexible physical techniques, including adapting to new scenarios and leveraging unconventional tools or environments.

Key Adaptations

- **Enhanced Situational Awareness:** Go beyond simply being aware of your surroundings; actively recognize potential threats and trust your instincts to avoid dangerous situations before they escalate.
- **Verbal De-escalation:** Learn to firmly and confidently set boundaries, communicate your intent to avoid conflict, and use calm assertiveness to de-escalate tense situations.

Flexible Physical Techniques:

- **Focus on escape:** Prioritize disengagement techniques to create distance or a barricade to escape a confrontation.
- **Adapt to the environment:** Practice using your surroundings for cover or as improvised defense tools, as this is often more practical than relying solely on standard moves.
- **Learn modern systems:** Study integrated self-defense systems like Krav Maga and Brazilian Jiu-Jitsu that teach practical, real-world techniques.
- **Mental & Emotional Preparedness:** Develop mental resilience to stay calm under pressure through controlled breathing and mental rehearsals, which allows for clear, decisive action.

CHAPTER 9

MEDICAL ASPECTS OF INCAPACITATION

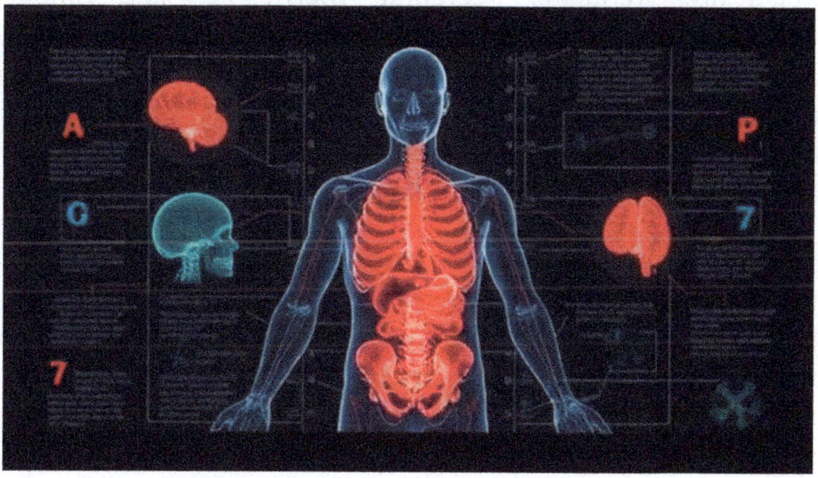

Immediate Medical Responses to Trauma

Upon sustaining acute trauma, especially to vital life support systems, the human body initiates an immediate, complex cascade of physiological responses. This initial shock, often manifesting as a sudden drop in blood pressure, loss of consciousness, or involuntary muscle spasms, is directly proportional to the impact's severity and location. Comprehending these instantaneous reactions is crucial for accurately assessing the efficacy of a targeted strike and predicting an adversary's immediate post-impact status. Such precise knowledge directly informs subsequent tactical decision-making in high-stakes confrontations.

Targeted trauma to the central nervous system, for instance, a direct strike to the brainstem or upper cervical spine, invariably triggers an immediate and profound systemic shutdown. This type of insult typically results in instantaneous loss of motor control, respiratory arrest, and a rapid collapse of circulatory function. Conversely, severe hemorrhaging from compromise to major arterial pathways precipitates swift hypovolemic shock, drastically reducing cerebral perfusion and inducing unconsciousness within a matter of seconds. Each critical system responds uniquely, yet often synergistically, to a severe traumatic event, leading to rapid incapacitation.

Long-Term Physiological Repercussions

Beyond the immediate cessation of hostile action, the physiological aftermath of precise targeting often leads to profound long-term repercussions. Injuries to the central nervous system, for instance, can result in irreversible neurological deficits, ranging from complete paralysis to severe sensory and motor impairments. Similarly, extensive skeletal trauma, particularly to major load-bearing joints or the spine, frequently culminates in chronic pain syndromes and significant mobility restrictions. Damage to vital internal organs, even if survivable in the short term, often precipitates chronic dysfunction, necessitating lifelong medical management and potentially organ replacement. These cascading effects underscore the permanent nature of the incapacitation sought by C.E.T.A. principles.

The systemic shock endured during a high-impact engagement can also initiate a cascade of secondary physiological complications that persist long after the initial trauma. Persistent circulatory issues, such as chronic hypertension or impaired vascular integrity, may develop, alongside compromised respiratory function from lung damage or thoracic impact. Such pervasive bodily insults invariably lead to a diminished quality of life, often requiring extensive rehabilitative therapy and ongoing pharmacological interventions. Understanding these enduring consequences is crucial for a comprehensive appreciation of the efficacy and finality inherent in calculated destruction.

Medical Documentation and Reporting

Accurate medical documentation following any engagement is not merely a bureaucratic exercise; it is a critical component of tactical analysis and legal accountability. Such records provide an objective baseline for understanding the immediate physiological effects of applied C.E.T.A. principles. This meticulous reporting allows for the validation or refinement of techniques and offers invaluable data for future training scenarios. Without precise documentation, the efficacy and consequences of an intervention remain largely anecdotal.

Effective reporting demands rigorous detail, focusing on observable trauma, specific anatomical impact points, and the immediate systemic responses. Each entry must clearly delineate the nature of the injury, its precise location, and any discernible changes in the subject's condition, such as respiratory distress or neurological impairment. This information forms the bedrock for forensic investigation, informs subsequent medical interventions, and contributes significantly to the ongoing development of C.E.T.A. protocols. Objectivity and factual accuracy are paramount in every recorded observation.

Forensic Analysis of Combat Injuries

When a physical engagement concludes, the injuries sustained by an individual offer a profound and often chilling narrative. Forensic analysis in this context moves beyond simple medical treatment, becoming a critical tool for dissecting the precise mechanics of incapacitation or neutralization. Every contusion, fracture, or internal trauma serves as a tangible record of the forces applied and the systems compromised. This examination allows for a detailed reconstruction of the encounter, revealing the efficacy of specific C.E.T.A. applications.

The systematic evaluation of combat injuries provides invaluable data regarding the precise targeting of the body's life support systems. For instance, skeletal fractures can indicate the direction and force of impact, while patterns of soft tissue damage might betray the type of weapon or technique employed. Examining neurotrauma or vascular disruption directly illustrates the success of attacks aimed at the Central Nervous System or Circulatory System. Such detailed forensic insight is crucial for understanding the physiological chain reactions initiated by calculated destruction.

Beyond mere documentation, the forensic analysis of combat injuries directly informs and refines tactical training methodologies. By meticulously studying the observable effects of various strikes and manipulations on the human form, practitioners can gain a deeper understanding of cause-and-effect relationships in high-stakes confrontations. This empirical feedback loop allows for the continuous optimization of C.E.T.A. principles, ensuring that techniques are not only theoretically sound but demonstrably effective in real-world applications. It transforms raw data into actionable intelligence for future preparedness.

CHAPTER 10

ADVANCED C.E.T.A. APPLICATIONS

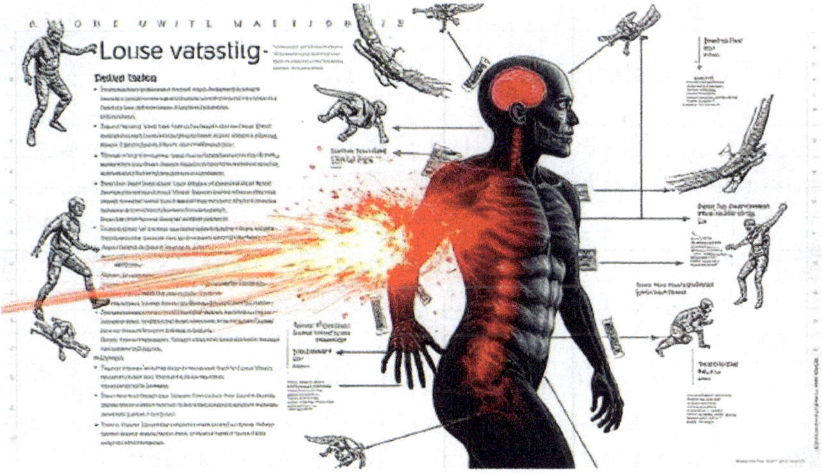

Close-Quarters Combat Strategies

Close-quarters combat (CQC) inherently demands a rapid, decisive application of force, making the principles of C.E.T.A. particularly relevant. In these compressed engagements, the objective shifts from merely neutralizing a threat to achieving immediate, irreversible incapacitation. Our focus remains on the precise targeting of critical life support systems, ensuring that every movement contributes to the swift cessation of hostile action. This approach minimizes exposure and maximizes the probability of success in high-stakes scenarios.

The unique spatial and temporal constraints of CQC necessitate a refined understanding of anatomical vulnerabilities under pressure. Reduced reaction times and minimal engagement distances amplify the importance of precise strikes to the Central Nervous System, major circulatory pathways, and the respiratory complex. Strategic application in these environments prioritizes efficiency, leveraging body mechanics and available angles to deliver the calculated destruction C.E.T.A. advocates. Mastery of these strategies transforms proximity from a limitation into a decisive advantage.

Weapon System Integration with C.E.T.A.

The core tenets of C.E.T.A. — the precise targeting of an adversary's vital life support systems — remain immutable, even with the introduction of a weapon. A weapon, whether bladed, impact, or ballistic, serves primarily as an extension, amplifying the destructive potential directed towards these critical physiological points. It transforms a combative engagement from a struggle of attrition into a swift, decisive act of incapacitation. Therefore, the strategic objective shifts from merely inflicting damage to ensuring immediate operational failure of the target through calculated application.

Integrating a weapon demands an even deeper understanding of anatomical vulnerabilities and the weapon's specific capabilities to exploit them. For bladed or impact tools, this means directing force to sever major arteries, collapse the respiratory system, or induce neurological shutdown with surgical precision. With firearms, the C.E.T.A. doctrine moves beyond generalized center mass engagements, advocating for precise shot placement to the central nervous system or major vascular structures for immediate cessation of hostile action. Ultimately, the weapon is merely a more efficient delivery mechanism for the calculated destruction C.E.T.A. prescribes.

Environmental Factors in Target Acquisition

C.E.T.A. principles, while anatomically precise, must always be applied within the context of dynamic operational environments. Seldom do engagements occur under ideal, controlled conditions; rather, they unfold amidst unpredictable variables. Factors such as lighting, ambient noise, structural obstacles, and surface stability significantly influence an operator's ability to identify, acquire, and effectively strike critical anatomical targets. A comprehensive understanding of these environmental interferences is paramount for successful target acquisition and neutralization.

Low-light conditions, for instance, can obscure visual cues necessary for precise targeting, necessitating reliance on proprioception and tactile feedback. Confined spaces or uneven terrain restrict an operator's movement and the trajectory of strikes, demanding adaptations in stance and limb extension. Similarly, high wind or rain can impair sensory perception and affect balance, both for the operator and the target, altering the dynamics of an engagement. Mastery of C.E.T.A., therefore, extends beyond anatomical knowledge to include the fluid adaptation of techniques to prevailing environmental challenges.

Scenario-Based Training and Drills

While theoretical comprehension of C.E.T.A. principles is foundational, true mastery demands rigorous application in dynamic environments. Scenario-based training serves as the critical bridge between academic understanding and practical execution. It immerses the practitioner in simulated high-stress situations, forcing the immediate recall and precise application of target acquisition methodologies. This type of training is indispensable for conditioning the necessary physiological and psychological responses under duress. Only through such practical immersion can the intricate science of calculated destruction be truly internalized.

Effective drills must meticulously replicate the unpredictable nature of real-world confrontations, incorporating varied threat profiles and environmental factors. The primary objective remains the rapid and decisive neutralization of threats by targeting the body's vital life support systems. Incorporating force-on-force exercises, even with protective gear, allows for the testing of decision-making processes and the immediate application of destructive force. Through consistent, high-intensity repetition, the C.E.T.A. methodology becomes an ingrained, instinctive response, rather than a conscious thought process. This engrains the critical actions required for immediate threat cessation.

VITAL POINTS Of THE HUMAN BODY

Vital points in the human body: essential organs and vulnerable pressure points

The human body possesses several vital organs that are indispensable for survival. These organs are the brain, heart, lungs, kidneys, and liver. Additionally, specific pressure points on the body, when impacted or stimulated, can have significant effects on the nervous system and potentially lead to severe pain, disability, or even loss of consciousness or death.

Brain:

The brain acts as the body's control center, responsible for thoughts, memories, emotions, movements, sensations, and regulating other organs' functions.

Heart:

Located in the chest, the heart pumps blood throughout the body, delivering oxygen and nutrients to cells and removing waste products.

Lungs:

These two spongy organs in the upper chest facilitate gas exchange, bringing oxygen into the blood and releasing carbon dioxide.

Kidneys:

Situated in the back of the abdomen, the kidneys filter blood, remove waste, form urine, and regulate blood pressure and red blood cell production.

Liver:

Located on the right side of the abdomen, the liver plays a critical role in metabolism, filtering blood, producing bile for digestion, and regulating blood clotting.

VUNERABLE POINTS OF THE BODY

Pressure points are sensitive areas on the body where nerves or other vital structures lie close to the surface. Targeting these points, particularly in the context of martial arts or self-defense, can amplify the impact of strikes or pressure, potentially causing pain, disorientation, or even unconsciousness.

Understanding the location and function of these organs is crucial for comprehending human biology and the importance of maintaining their health.

HIGH SECTION

Eyes

Entry points into the brain, blindness.

Temple

Knockout, point, dizziness, unconsciousness, brain trauma

Forehead

Knockout point, dizziness, brain trauma

Under Nose

Crushed nasal bone, Watery eyes, hemorrhage

Bridge of Nose

Broken nose, watery eyes.

Cheekbone

Crushed Cheekbone.

Jaw

Knockout point, broken jaw, unconsciousness.

Ears

Deafness, disorientation, dizziness, loss of balance.

Philtrum

Located above upper lip at mustache line, broken top row of teeth.

Throat

Damaged trachea, internal swelling of throat, Adam's apple crush.

Neck

Broken neck, paralysis.

Clavicle

Extreme pain, broken collar bone, immobilization of both arms.

Clavicle-Notch

Pain point located Behind clavicle bone. Knife access to subclavian artery.

Hypoglossal

Pain point located underneath jawline.

Carotid Artery

Unconsciousness, excessive bleeding, death.

Brachial Plexus

Pain point, Knockout point.

Brain Stem

Located at base of the skull, (Knife Blade) immediate immobilization and or death. Severe damage to the medulla oblongata.

MID SECTION

Sternum

Crushed chest cavity, heart stoppage, lung damage.

Xiphoid Process.

Located where the rib cage meets the bottom of the

Sternum

The snapping off the xyphoid process can lead to internal bleeding and or death.

Bicep Muscle -

A direct hit to the bicep muscle can immobilize the arm.

Bicep Notch

Pain point located at the bottom of the bicep muscle.

Inside of forearm

Pain point located at middle of forearm.

Radial Nerve

Located just below the insertion of the deltoid muscle.

Pictorial Muscle

Pain point. Located at center of pictorial, chest muscle.

Deltoid Muscle

Located at shoulder.

Solar Plexus

Located near the typhoid process at the center of the body's front, where the rib cage meets the bottom of the sternum.

Stomach

A direct punch, kick, or stomp here can easily knock the wind out of An attacker.

Oblique Skin Pinch

A great deal of pain can be generated by placing both open hands on the obliques and squeezing them.

Ribs

The ribs can be easily broken using straight, hook, or uppercut punches, kicks, and elbow strikes.

Kidneys

Attack the kidneys using a knife, rifle, bayonet, hook, punch, or front thrust kick.

Spinal Column

The spine can be attacked using a punch, elbow, kick, stomp, and knife blade. It can cause immediate paralysis if damaged.

Coccyx

Or tailbone. It can cause paralysis of the lower extremities if damaged.

LOW SECTION

Groin

Extremely vulnerable and open to attack. There is no outer protection to protect the groin area. The groin can be attacked using kicks, strikes, or knife attack. The testicles are the primary target.

Knee

The knee is one of the strongest joints in the body, yet one of the weakest at the same time. Because the knee joint supports most of the weight of the body, a great deal of stress is placed on the knee joints. The knee can be broken. Quite easily using kicks, joint manipulation, or palm heel strikes.

Inside of Knee

The inside of the knee can be easily attacked from behind the attacker with a side. Front kick, low round kick, hook punch, or knife slash. Cutting the tendons can easily render the leg useless.

Side of Knee

Because the knee bends in only One Direction, backwards and forward, A thrust kick to the side of the knee can break it instantly. This injury causes extreme pain and instantly immobilizes the attacker's leg.

Shin

Or, better known as the tibia, is a very sensitive and tender area for most people. A strike here can cause extreme pain or mobilize the leg.

Ankle

The ankle can be broken by twisting sharply.

Achilles Tendon

This highly sensitive target area can quickly bring an attacker to his knees. Attack this target using a low line front kick, sweeping heel kick, or knife slash to completely immobilize.

Instep

The end step is an excellent target to attack if grabbed from behind in a bear hug. Type position, raise your knee. Up high and stomp. The attacker's in step with the heel of your foot, boot heel, etcetera. Note: Do not stomp the toes as this will only aggravate the attacker More and will not produce the desired effect. Attacking the instep will break the small bones, tarsals, and metatarsals of the foot, quickly immobilizing the attacker.

Pelvic Triangle

The pelvic triangle is a region in the lower part of the pelvis that is bounded by the pubic symphysis anteriorly, the ischial tuberosities laterally, and the sacrum posteriorly. The hip joint which is a ball-in-socket joint can be dislocated by directly targeted kick or gun shot to this joint. Any gun shot to the pelvic triangle in general will drop the attacker instantly in his tracks, as this structure supports the body through the hip joints and legs.

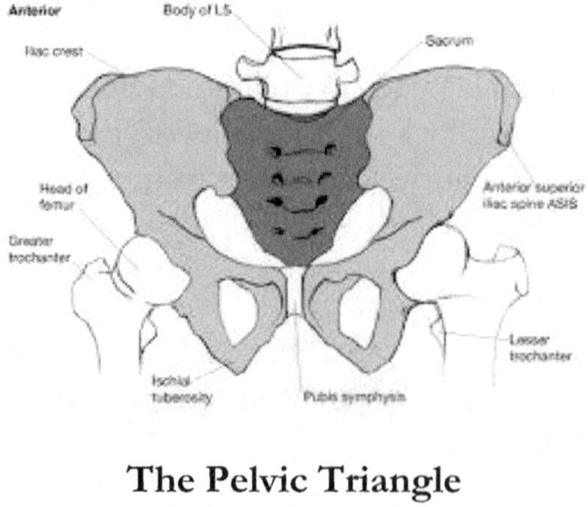

The Pelvic Triangle

CHAPTER 11

CONCLUSION AND FUTURE PERSPECTIVES

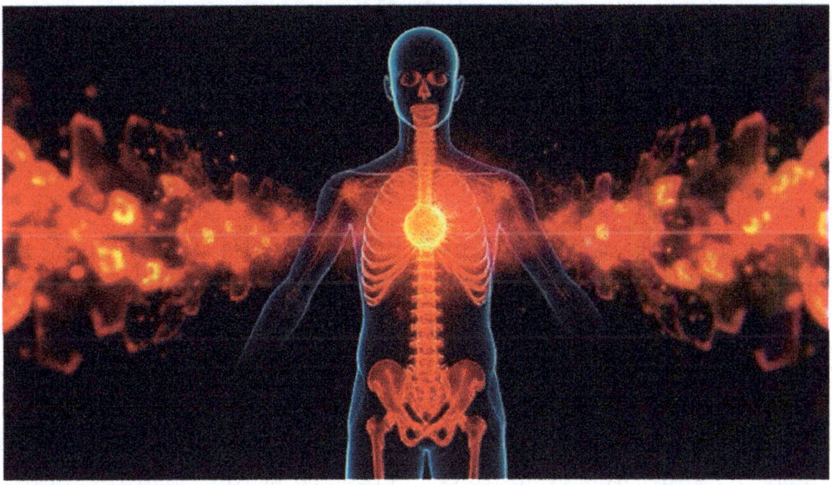

Synopsis of C.E.T.A. Principles

Throughout this text, we have systematically dissected the human body to understand its inherent vulnerabilities. C.E.T.A., or Close-Engagement-Target-Acquisition, was conceived from the imperative to neutralize threats with unparalleled speed and efficiency. Its fundamental principle lies in the calculated disruption of an adversary's critical life support systems, ensuring a definitive conclusion to any physical confrontation. This approach moves beyond mere incapacitation, aiming for a complete cessation of hostile action through precise physiological targeting.

The core of C.E.T.A.'s methodology revolves around an in-depth understanding of the Central Nervous System, Circulatory System, Respiratory System, and the Brain, alongside the critical components of the Skeletal System and internal organs. Each chapter meticulously detailed the anatomical targets within these systems, explaining the precise physiological consequences of various impacts. This comprehensive analysis provides the operator with the knowledge necessary to exploit these weaknesses, transforming potential chaos into controlled, decisive action. Such mastery is paramount for professionals operating in high-stakes environments where immediate threat neutralization is non-negotiable.

Continuous Professional Development

Mastery in any tactical discipline is never a static achievement; it is a relentless pursuit. The principles laid out in C.E.T.A. provide a robust foundation, yet the dynamic nature of real-world engagements demands continuous refinement of one's skills. Professionals operating in high-stakes environments must consistently revisit and practice these core concepts to ensure peak operational readiness. This ongoing commitment to learning is not merely advisable, but absolutely critical for effective threat neutralization.

True professional development extends beyond physical drills; it encompasses a deep, evolving understanding of the human physiological landscape. New research and practical experience continuously inform our insights into the body's vulnerabilities and reactions under duress. Therefore, integrating new knowledge, analyzing past encounters, and adapting methodologies are vital for maintaining an edge. Staying current ensures that the application of C.E.T.A. remains maximally effective in diverse scenarios.

The Evolving Landscape of Tactical Engagement

The landscape of tactical engagement is in a state of perpetual flux, driven by advancements in technology, shifts in geopolitical dynamics, and the unpredictable nature of human conflict. While the fundamental vulnerabilities of the human body, as detailed throughout this text, remain constant, the contexts in which these weaknesses are exploited are continually evolving. This necessitates a dynamic approach to training and a proactive mindset in adapting C.E.T.A. principles to novel threats and operational environments.

This evolution isn't solely confined to the development of new weaponry; it encompasses changing doctrines, the increasing prevalence of urban combat scenarios, and even the psychological profiles of adversaries. Professionals must therefore remain vigilant, integrating new intelligence and practical experience into their understanding of close-engagement scenarios. The principles of C.E.T.A. provide an adaptable framework, but their effective application demands constant refinement against contemporary challenges.

The emphasis on precision targeting of life support systems becomes even more critical in complex, rapidly developing scenarios. As operational environments become more unpredictable and variables multiply, the ability to quickly and decisively neutralize a threat with minimal collateral impact is paramount. This necessitates a deep, intuitive understanding of human anatomy and immediate physiological responses. Ultimately, this underscores the enduring relevance of C.E.T.A.'s core philosophy, ensuring mission success and operator safety amidst an ever-changing threat matrix.

Final Remarks on Responsible Application

The insights presented within this text, detailing the calculated destruction of the human body's life support systems, carry an immense weight of responsibility. This knowledge, while critical for professionals operating in high-threat environments, is not to be taken lightly or applied indiscriminately. Its mastery demands not only technical proficiency but also a rigorous ethical framework that dictates its use solely in situations of extreme necessity. Understanding vulnerability necessitates an equal understanding of judicious restraint.

As emphasized throughout, C.E.T.A. is a system developed for academic study and professional development, focusing on the tactical incapacitation of threats when all other options have been exhausted. It is a precise science intended for the most critical moments, not for casual application or malicious intent. Therefore, the onus of responsible and lawful deployment of these principles rests entirely upon the individual, ensuring this profound understanding serves only to protect and preserve life within defined operational parameters.

THANK YOU

Special Thanks To:

GEN. Gordon Sullivan (U.S. Army Chief of Staff)
GEN. Colin Powell (Chairman of the Joint Chiefs of Staff)

SMA Richard A. Kidd (Sergeant Major of the Army)
SMA Michael A. Grinston (Sergeant Major of the Army)

CSM. Lesley Freeman Jr.
CSM. James McCall
CSM. William Perry
CSM. Richard Young
CSM. Edward Deane
CSM Devon Kingston
CSM. Stanley Mitchell
CSM. Larry Dinkle
CSM. Andrew McFowler
CSM. Larry Jordan, (Special Forces)
CSM. Lawrence Henry, (Special Forces)
CSM. James E. Turner, Jr. (Special Forces)

COL. Dennis R. Lewis
COL. Thomas Ellis
COL. Scott Marcy
COL. Steven Sukuma
COL. John Dominick
COL. Ronald Jones
COL. Thomas Costello
COL. John Braden, Jr.
COL. Edward Broderick

COL. Charles McPherson

LTC. Thomas Keller
LTC. Alexander Smith
LTC. Arnaldo Claudio
LTC. Randy Garibay
LTC. Charles Childers
LTC. Robert Madden
LTC. Gerald Swenson

MAJ. Mark E. Rambis
MAJ. Jose M. Marrero
MAJ. Oliver Jenkins

CPT. Brian Maka
CPT. Brian T. Boyle
CPT. Mark A. Moser
CPT. Barry V. Hadley
CPT. Randall A. Bartlett

SR. DS. Himes
DS. Gaskins
Basic Training Drill Instructors

Judge Roy S. Moore
Chief Justice, Alabama Supreme Court

David Dreier
United States Congressman, California

Alice Edwards
Sr. Editor, Army Trainer Magazine

Richard Hinson
Advanced Law Enforcement Division, MP School

Joe Jennings
President, Panther Productions

Dr. Royce Jones, DC.
Chiropractor

Dr. E.A. Moore, Ph.D., MD.
President/Founder
World Martial Arts Hall of Fame

Assistant CQC Instructors (U.S. Army)
James E. Turner, Jr.
Maceo Jordan
Hence Spencer
Donald Mouton
Andrew Hewuse
Michael Maxwell
Irving Domenech

About The Author

Dr. Leonard C. Holifield, DMS

Dr. Leonard Holifield is Director and Founder of the U.S. Educational Institute Phalanx Training Academy, LLC, (I.A.E.P.A) International Academy of Executive Protection Agents, (I.A.M.S.P) International Academy of Martial Science and Philosophy, (I.G.S.C) International Grandmasters Sokeship Council, (SKF) Sikaron Karate Federation, and the (P.E.H.F) Phalanx Elite Hall of Fame.

With over 40 years of combined experience in the fields of Security, Executive Protection, Military, Law Enforcement, and Education, he is highly sought after for his expertise in Executive Protection and real-world Close-Quarter-Combat.

The former decorated Combat Veteran of the Gulf War – was credited with capturing the first Iraqi enemy POW (Prisoner of War) during Operation Desert Storm.

As an honor graduate and former instructor for the prestigious ESI – Executive Security International bodyguard academy, he went on to become the Executive Security Officer and Personal Bodyguard to the former Chief Justice of the Alabama Supreme Court, Chief Justice Roy Moore, aka "The Ten Commandments Judge" from 2000 to 2007. He has protected High Profile corporate CEO's, Government Officials, Movie and Music Celebrities, Executives, and Foreign Dignitaries.

A graduate of the World Academy of Martial Arts Philosophy and Science, Malaysia, he is one of only (10) Americans from the United States to earn the coveted Doctor of Martial Science DMS Degree from this prestigious institution.

5X inductee into the World Martial Arts Hall of Fame, Recipient of the Who's Who in America 2004 and Who's Who in the World 2005, he is considered one of the leading experts in the United States in Military Combatives.

During his military career, he served as Chief Combatives Instructor to the United States Army and was instrumental in rewriting the Army's field manual on hand-to-hand combat (FM 21-150), Combatives, December 1971.

His positions included NBC/NCO Nuclear-Biological-Chemical Specialist, Combat Lifesaver, Master Fitness Instructor, Instructor Certified "H" Identifier, MP Inmate Labor Supervisor, Tac-Fire Operations Specialist (13C), Military Police Officer (95B), and CQC trainer to the FATC Field Artillery Training Center, Ft Sill, Ok, 101

Airborne, Air Assault, Air Calvary, Infantry, (SRT) Special Reaction Teams, Delta, and Special Forces.

His work in the field of hand-to-hand combat has been featured on AFKN (Armed Forces Korean News Network), Stars & Stripes, Soldiers, Army Trainer, Korus, Military Police, and Black Belt Magazines. His work has been recognized by top military leaders, including former Chairman of the Joint Chiefs of Staff, Gen. Collin Powell, Sergeant Majors of the Army Richard Kidd, Michael Grinston, Command Sergeant Major Leslie Freeman, FORSCOM Command Sergeant Major Todd Sims, Col. Dennis Lewis, and Former Army Chief of Staff Gen. Gordon Sullivan.

MILITARY HONORS EARNED

Army Achievement Medal (4), Army Commendation Medal (6), Army Commendation Medal w/Valor (1), Good Conduct Medal (3), Southwest Asia Service Medal, Saudi Arabian Kuwait Liberation Medal, Soldiers Medal, Meritorious Service Ribbon, Army NCO Professional Development Ribbon, National Defense Ribbon, Army Overseas Ribbon, Soldier of the Month, Battalion Certificate of Achievement in Hand-to-Hand Combat, JROTC Certificate of Meritorious Service, Saudi Arabian Liberation of Kuwait Medal.

AUTHORSHIP & PUBLICATIONS

SOLO, Security Operations for the Lone Officer, Outskirts Press, 2007

Close Quarter Combat – A Soldiers Guide to Hand-to-Hand Combat, Paladin Press, 1997

Army Trainer Magazine - Sometimes the Enemy is Close Enough to Touch, 1992 Army Trainer Magazine - Hand-to-Hand Combat in Today's Army, Cover Story

Military Police Magazine - USD – Are You Really Ready? 1996

Soldiers Magazine, United States Army (Profile) 1990

Soldiers Magazine, Hand-to-Hand Training, 1995

KORUS Korean Magazine - Fist of Steel, 1993

CETA, Close Engagement Target Acquisition, by Leonard C. Holifield, 2025

Armageddon – The Total Destruction of the Arms, 2025

The Art of Sikaron Karate, 2025

The SOLO Protector, 2025

Video Artist for Panther Productions, Close Quarter Fighting Series, 1995 Military Combatives Masters of the 20th Century, Lulu Press, 2006.

Phalanx Training Academy, LLC

110 Walter Way 2522

Stockbridge, Ga. 30281-2522

Phone-1: (352) 718-0995

Phone-2: (770) 742-5269

Email: Phalanxeps@aol.com

Website: www.Phalanxta.com

Made in the USA
Monee, IL
03 November 2026

67375813R00066